CDMA RF System Engineering

For a complete listing of the *Artech House Mobile Communications Library,*
turn to the back of this book.

CDMA RF System Engineering

Samuel C. Yang

Artech House
Boston • London

Library of Congress Cataloging-in-Publication Data
Yang, Samuel C.
    CDMA RF system engineering / Samuel C. Yang
        p.   cm. — (Artech House mobile communications library)
    Includes bibliographical references and index.
    ISBN 0-89006-991-3 (alk. paper)
    1. Wireless communication systems.   2. Code division multiple access.
    3. Personal communication service systems.   I. Title.   II. Series.
    TK5103.2.Y36   1998
    621.3845—dc21   98-10451
                                                                            CIP

British Library Cataloguing in Publication Data
Yang, Samuel C.
    CDMA RF system engineering — (Artech House mobile communications library)
    1. Code division multiple access
    I. Title
    621.3'845

    ISBN 0-89006-991-3

Cover design by Nina Y. Hsiao

International Standard Book Number: 0-89006-991-3
Library of Congress Catalog Card Number: 98-10451

10 9 8 7 6 5 4 3 2 1

*To my loving parents John and Hannah,
my caring sisters Esther and Nina,
and my precious wife Jenny*

Contents

Preface   *xv*

Acknowledgments   *xix*

1   Introduction   **1**

1.1   Motivation   1

1.2   Multiple Access Using Spread Spectrum   2

1.3   Applications of DS-SS in Mobile Communication   9

2   Radio Propagation   **13**

2.1   Link Analysis   13

2.2   Propagation Loss   14

2.2.1 Free-Space Model   15

2.2.2 Lee Model   15

2.2.3 Hata Model   16

2.2.4 Observations   17

2.3   Shadowing   19

2.4   Multipath Rayleigh Fading   19

2.5   Multipath Delay Spread   23

3	**Fundamentals of Digital RF Communication**	29
3.1	Introduction	29
3.2	System Components	31
3.3	Source Coding	32
3.3.1	Characteristics of Human Speech	33
3.3.2	Vocoders	34
3.4	Channel Coding	36
3.4.1	Linear Block Codes	37
3.4.2	Convolutional Codes	41
3.4.3	Interleaving	43
3.5	Multiple Access	43
3.5.1	Walsh Codes	46
3.5.2	PN Codes	51
3.6	Modulation	58
3.6.1	Binary Phase-Shift Keying (BPSK)	58
3.6.2	Quadrature Phase-Shift Keying (QPSK)	66
3.6.3	Applications in IS-95 CDMA System	72
4	**Principles of Code Division Multiple Access**	75
4.1	Introduction	75
4.2	Capacity	75
4.2.1	Effects of Loading	78
4.2.2	Effects of Sectorization	79
4.2.3	Effects of Voice Activity	82
4.3	Power Control	83
4.3.1	Why Power Control?	83
4.3.2	Reverse Link	85
4.3.3	Forward Link	94
4.4	Handoff	94
4.4.1	Set Maintenance	99

4.4.2	Handoff Process	100
4.4.3	Pilot Search	102
5	**Link Structure**	**105**
5.1	Asymmetric Links	105
5.2	Forward Link	105
5.2.1	Pilot Channel	106
5.2.2	Sync Channel	106
5.2.3	Paging Channel	109
5.2.4	Traffic Channel	114
5.2.5	Modulator	118
5.3	Reverse Link	118
5.3.1	Access Channel	119
5.3.2	Traffic Channel	122
5.4	Traffic Channel Formats	125
5.4.1	Forward Link	128
5.4.2	Reverse Link	130
6	**Call Processing**	**133**
6.1	Call Processing States	133
6.2	Initialization State	135
6.2.1	System Determination Substate	135
6.2.2	Pilot Channel Acquisition Substate	135
6.2.3	Sync Channel Acquisition Substate	135
6.2.4	Timing Change Substate	136
6.3	Idle State	136
6.3.1	Paging Channel Monitoring	136
6.3.2	Idle Handoff	137
6.3.3	Paging Channel Messages	137
6.4	Access State	139
6.4.1	Update Overhead Information Substate	139

6.4.2   Page Response Substate   140

6.4.3   Mobile Station Origination Attempt Substate   140

6.4.4   Registration Access Substate   140

6.4.5   Mobile Station Order/Message Response Substate   140

6.4.6   Mobile Station Message Transmission Substate   141

6.4.7   Access Procedures   141

6.5   Traffic Channel State   145

6.5.1   Traffic Channel Initialization Substate   145

6.5.2   Waiting for Order Substate   146

6.5.3   Waiting for Mobile Station Answer Substate   146

6.5.4   Conversation Substate   147

6.5.5   Release Substate   147

7   CDMA Design Engineering   149

7.1   Introduction   149

7.2   Forward Link Analysis   149

7.2.1   Pilot Channel   149

7.2.2   Traffic Channel   156

7.3   Reverse Link   158

7.3.1   Traffic Channel   159

7.3.2   Reverse-Link Rise   163

7.3.3   Frequency Reuse Factor   164

7.4   PN Offset Planning   165

7.4.1   Short PN Sequences   165

7.4.2   Co-PN Offset   168

7.4.3   Adjacent PN Offset   171

7.5   9.6-Kbps and 14.4-Kbps Systems   174

7.5.1   Voice Quality   174

7.5.2   Power Control—Forward Link   175

7.5.3   Coverage   176

7.5.4   Capacity   179

8	**CDMA Performance Engineering**	**181**
8.1	Introduction	181
8.2	Channel Supervision	181
8.2.1	Forward Link	181
8.2.2	Reverse Link	182
8.3	Power-Control Parameters	182
8.4	Search-Window Sizes	184
8.4.1	SRCH_WIN_A	184
8.4.2	SRCH_WIN_N and SRCH_WIN_R	187
8.5	Field Optimization	189
8.5.1	Pilot Strength	190
8.5.2	FER	190
8.5.3	Forward Link Coverage	190
8.5.4	Forward-Link Interference	191
8.5.5	Reverse-Link Coverage	192
8.5.6	Reverse-Link Interference	192
8.5.7	Some Concluding Remarks	193
9	**System Noise Management**	**195**
9.1	Introduction	195
9.2	Types of Interference	196
9.2.1	Forward Link	196
9.2.2	Reverse Link	197
9.3	Thermal Noise	198
9.4	Low-Noise Amplifier	199
9.4.1	Baseline System Without LNAs	200
9.4.2	System With LNA	202
9.4.3	Signal-to-Noise Ratio Improvement	205
9.4.4	Capacity Improvement	208
9.5	Intermodulation	208

9.5.1	Intermodulation Theory	208
9.5.2	CDMA Scenario	213
9.6	Interference Due to Other Mobiles	215
10	**CDMA Traffic Engineering**	**217**
10.1	Introduction	217
10.2	Fundamental Concepts	218
10.2.1	Traffic Intensity	218
10.2.2	Loads	220
10.3	Grade of Service	220
10.3.1	Erlang-B Model	221
10.3.2	Erlang-C Model	223
10.4	CDMA Applications	225
10.4.1	Soft Blocking	225
10.4.2	Hard Blocking	231
11	**Management Information Systems for Personal Communication Networks**	**235**
11.1	Introduction	235
11.2	Management Information Systems	236
11.2.1	Information System and Control	236
11.2.2	Classes of Decisions	238
11.3	Network Management	240
11.3.1	Fault Management	240
11.3.2	Performance Management	242
11.3.3	Configuration Management	243
11.3.4	Planning	244
11.3.5	Call Accounting	245
11.4	Concluding Remarks	246
12	**RF Regulatory Considerations**	**249**
12.1	Motivation	249

12.2	SAB Determination	250
12.2.1	Review of AMPS SAB Calculation	250
12.2.2	CDMA SAB Determination With Multiple Sectors	253
12.2.3	CDMA SAB Determination With Single Sector	257
12.2.4	CDMA SAB Determination With Power Spectral Density	260
12.3	RF Exposure Rules	262
12.3.1	Maximum MPE Limits	263
12.3.2	Application of MPE Limits	264
12.3.3	Evaluation of MPE Power Densities	267
12.3.4	RF Mitigation Measures	269
12.4	Remarks	269
	About the Author	**271**
	Index	**273**

Preface

The wireless communications industry has been undergoing tremendous changes in the last few years. With the auction of *personal communication services* (PCS) licenses in the United States, most incumbent service providers found themselves competing with not just one, but several other service providers who offered comparable services at competitive prices. At the same time, the wireless subscriber base has been increasing as well, with some projecting the total number of worldwide wireless subscribers reaching over 360 million by year 2000.

The tremendous market growth coupled with fierce competition implies that each service provider must differentiate itself from the competitors by offering a *high-quality* service at a *competitive* price. From an engineering perspective, the first goal may be attained by optimally designing and maintaining the network such that the customer's calling experience nearly replicates that of a landline phone. The second goal may be achieved by effectively and efficiently planning, managing, and operating network resources. For many service providers, *code division multiple access* (CDMA) manifested in the form of a IS-95 wideband spread-spectrum system has played a key role in achieving both goals. Many technical features of CDMA, which this book describes in detail, enable the network to offer high-quality on-demand voice services to customers. At the same time, CDMA's ability to provide high capacity allows a service provider to better utilize its invested network assets, lower its cost structure, and thus lower its service pricing.

In an effort to provide *radio frequency* (RF) and system engineers with the ability to optimally engineer and manage an IS-95 based network as well as to provide students with an inclusive treatment of spread-spectrum technology, this book has been written to give a comprehensive coverage of CDMA RF system engineering. The book emphasizes both theoretical and application aspects of code division as specifically applied to engineering a land-mobile network. The intended audience is practicing engineers and managers, senior-level undergraduates, and first-year graduate students.

To the extent possible, the relationship between general areas of digital communication and specific features of IS-95 is emphasized in the book. Other areas of land-mobile communications engineering, such as network management and traffic engineering, are also treated, with an emphasis on CDMA application. Furthermore, the chapters are modularized so the readers can read only those sections that are relevant to his or her needs. The book develops the idea of CDMA communication in the context of a land-mobile wireless network. To that end, the book is organized as follows.

Chapter 1 starts with a brief introduction of multiple access using direct-sequence spread-spectrum techniques. Multiple access is illustrated with the use of orthogonal codes, and some inherent benefits and difficulties of direct-sequence spread spectrum in a mobile communications environment are addressed. Chapter 2 reviews radio propagation from the perspectives of static and dynamic effects (i.e., path loss as well as shadowing and multipath phenomena).

The material on communication engineering of a CDMA network begins in Chapter 3 with a review of the fundamentals of digital communications; the chapter emphasizes only those aspects of digital communication applicable to an IS-95 based system. Chapter 4 introduces and describes the fundamental and theoretical concepts of spread-spectrum communication, while Chapters 5 and 6 describe the channel structure and call processing functions of an IS-95 based system. These three chapters serve as the background and foundation leading into the chapters that follow: Chapters 7 and 8 cover the essential materials of design and performance engineering of a CDMA network.

In migrating from an AMPS to a CDMA system, the cellular engineering paradigm effectively shifts from frequency planning to noise management, since every decibel of in-band noise reduced translates into capacity and coverage gains. The goal of Chapter 9 is to cover those special areas to which RF and system engineers should pay special attention in order to reduce in-band noise. Chapters 10, 11, and 12 contain special topics relevant to the operation and management of a CDMA network, such as traffic engineering, network management, and regulatory compliance issues.

At this point, a few words about the design of this book's cover are probably in order. The cover is an illustration of four superimposed layers, each representing a different aspect of CDMA technology. The first layer is a rigid matrix of hexagons which symbolizes the conventional analog cellular technology. The second layer depicts the technical aspects of CDMA-important equations and operating frequencies, the third layer shows 14 hexagonal volumes portraying breathing CDMA cells with different capacities.

Samuel C. Yang
Irvine, California

Acknowledgments

It is impossible to acknowledge all those people who have had a major influence on the conception and fruition of this book. To the best of my ability I shall attempt to do so. I would like to thank William C. Y. Lee and Dr. David Lee at AirTouch Communications for reviewing and approving the manuscript for publication. I would also like to thank Fernando Rico, Alix Watson, and Dr. Jin Yang, who have reviewed and provided valuable suggestions on parts of the manuscript. Dr. Jin Yang and Derek Bao have tirelessly answered many of my questions regarding the implementation details of an IS-95 system. Special thanks to Professor Lorne Olfman, who provided important reviews of parts of the manuscript. Furthermore, I would like to express my sincere appreciation to the special group of RF and traffic engineers that I work with and who challenge me every day on the engineering and operational details of a large and complex CDMA network.

My gratitude also goes to my sister Nina Y. Hsiao, who conceived the cover design for this book. I am so very thankful for all the effort that she put into the design. Her care for me and labor, as well as her unparalleled creativity, are sincerely appreciated. I would also like to thank my brother-in-law Howell Hsiao, Principal of Envision in Mountain View, California, for lending his unhesitant support throughout the cover design project.

In closing, I want to thank the most important participant in the writing of this book, my wife Jenny. This book would not have been possible without her unselfish love, support, and understanding during the many months of writing. She has endured my frustrations and shared in my delights. For her quiet and loving participation, I am so very much grateful.

1

Introduction

1.1 Motivation

The market of wireless communication is expected to increase dramatically in the late 1990s and beyond, and this expected high demand has led many service providers to investigate digital technology to satisfy the increasing demand. Spread spectrum has been used for a long time in military communications to resist intentional jamming and to achieve low-probability of detection. However, in recent years, spread spectrum has moved from military to commercial communications, culminating in the introduction of Interface Standard IS-95 *code division multiple access* (CDMA) technology as an alternative standard for commercial digital cellular and *personal communication system* (PCS) networks. Service providers, both cellular and PCS carriers, have deployed commercial CDMA systems in major metropolitan areas. The IS-95 CDMA is now being used in numerous cellular and PCS markets around the world. Service providers are deploying these systems in their markets, where there are mounting demands for higher capacity.

Multiple access systems share a fixed resource (i.e., frequency spectrum) to provide voice channels on demand to users. At first it seems counter-intuitive that deliberately increasing the bandwidth required for transmission increases capacity. After all, in a traditional *frequency division multiple access* (FDMA) scheme, increasing the required bandwidth per user decreases the total number of users a fixed spectrum can support. We start with a definition of spread spectrum [1]:

Spread-spectrum is a means of transmission in which the signal occupies a bandwidth in excess of the minimum necessary to send the information; the band spread is accomplished by means of a code that is independent of the data, and a synchronized reception with the code at the receiver is used for despreading and subsequent data recovery.

This book primarily deals with a type of spread spectrum that is employed in the IS-95 standard called *direct-sequence spread spectrum* (DS-SS). Another form of spread spectrum is called *frequency-hopping spread spectrum* (FH-SS) where the carrier frequency of the signal is moved (hopped) around in the band in a pseudorandom fashion. The result is an increase in effective bandwidth over time [2].

1.2 Multiple Access Using Spread Spectrum

Traditional ways of separating signals in time (i.e., *time division multiple access,* (TDMA)), or in frequency (i.e., FDMA) are relatively simple ways of making sure that the signals are orthogonal and noninterfering. However, in CDMA, different users occupy the same bandwidth at the same time, but are separated from each other via the use of a set of orthogonal waveforms, sequences, or codes. Two real-valued waveforms x and y are said to be orthogonal if their *cross-correlation* $R_{xy}(0)$ over T is zero, where

$$R_{xy}(0) = \int_0^T x(t)y(t)dt \tag{1.1}$$

In discrete time, the two sequences **x** and **y** are orthogonal if their *cross-product* $R_{xy}(0)$ is zero. The cross product is defined as

$$R_{xy}(0) = \mathbf{x}^T \mathbf{y} = \sum_{i=1}^{I} x_i y_i \tag{1.2}$$

where

$$\mathbf{x}^T = \begin{bmatrix} x_1 & x_2 & \cdots & x_I \end{bmatrix}$$

$$\mathbf{y}^T = \begin{bmatrix} y_1 & y_2 & \cdots & y_I \end{bmatrix}$$

Note that T denotes the transpose of the column vector, which is another representation of a sequence of numbers. For example, the following two sequences or codes, **x** and **y**, are orthogonal:

$$\mathbf{x}^T = \begin{bmatrix} -1 & -1 & 1 & 1 \end{bmatrix}$$

$$\mathbf{y}^T = \begin{bmatrix} -1 & 1 & 1 & -1 \end{bmatrix}$$

because their cross-correlation is zero; that is,

$$R_{xy}(0) = \mathbf{x}^T \mathbf{y} = (-1)(-1) + (-1)(1) + (1)(1) + (1)(-1) = 0$$

In order for the set of codes to be used in a multiple access scheme, we need two additional properties. In addition to the zero cross-correlation property, each code in the set of orthogonal codes must have an equal number of 1s and −1s [3]. This second property gives that particular code the pseudorandom nature. The third property is that the *dot product* of each code scaled by the order of the code must equal to 1. The order of the code is effectively the length of the code, and the dot product is defined as a scalar obtained by multiplying the sequence by itself and summing the individual terms; that is, the dot product of the code **x** is

$$R_{xx}(0) = \mathbf{x}^T \mathbf{x} = \sum_{i=1}^{I} x_i x_i \qquad (1.3)$$

The two orthogonal codes in the previous example also satisfy the second and the third conditions. Both **x** and **y** have an equal number of 1s and −1s, and the scaled dot products are

$$\left(\mathbf{x}^T \mathbf{x}\right)/4 = (-1)(-1) + (-1)(-1) + (1)(1) + (1)(1) = 4/4 = 1$$

$$\left(\mathbf{y}^T \mathbf{y}\right)/4 = (-1)(-1) + (1)(1) + (1)(1) + (-1)(-1) = 4/4 = 1$$

Note that the order of each code is 4.

Here, we summarize the properties of the set of orthogonal codes to be used in DS-SS multiple access:

1. The cross-correlation should be zero or very small.
2. Each sequence in the set has an equal number of 1s and −1s, or the number of 1s differs from the number of −1s by at most 1.
3. The scaled dot product of each code should be equal to 1.

Figure 1.1 illustrates the principle of a DS-SS multiple access scheme. Although these systems are often used for digital communication, we show their continuous-time equivalent in order to illustrate operating principles. We show two users simultaneously transmitting two separate messages, $m_1(t)$ and $m_2(t)$, in the same frequency band at the same time. The two users are separated from each other via the multiplication of orthogonal codes $c_1(t)$ and $c_2(t)$, which are the continuous-time versions of the two orthogonal codes $\mathbf{x}$ and $\mathbf{y}$ mentioned previously. Message $m_1(t)$ is multiplied by the code $c_1(t)$, and message $m_2(t)$ is multiplied by the code $c_2(t)$. The resulting products are added together by the adder and transmitted through the channel. In this case, we assume perfect synchronization of the codes at the receiver. If there are negligible errors over the channel, the recovered messages $\tilde{m}_1(t)$ and $\tilde{m}_2(t)$ will match the original messages $m_1(t)$ and $m_2(t)$ perfectly. In this example, we are

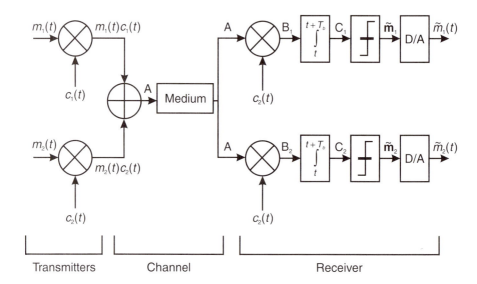

Figure 1.1 An example showing the operating principle of DS-SS multiple access. Two users are sending two separate messages, $m_1(t)$ and $m_2(t)$, simultaneously through the same channel in the same frequency band and at the same time. Through the use of orthogonal codes $c_1(t)$ and $c_2(t)$, the receiver recovers the two messages perfectly.

interested in sending two separate messages: $\mathbf{m}_1$, which is $(+1, -1, +1)$, and $\mathbf{m}_2$, which is $(+1, +1, -1)$.

Figure 1.2 shows the waveforms and spectrums for the two messages $m_1(t)$ and $m_2(t)$, the two orthogonal codes $c_1(t)$ and $c_2(t)$, and the two spread messages $m_1(t)c_1(t)$ and $m_2(t)c_2(t)$. While we do not go into the details of calculating the spectrums of these time waveforms, it suffices for our purposes to state that the bandwidth of a random digital waveform is limited to $1/T$, where T is the bit interval of the random digital waveform. We further make the distinction between T_b and T_c, where T_b is the *bit* interval (in seconds) of the message and T_c is the *chip* of the running orthogonal code. In this example, the chip rate $(1/T_c)$ of the orthogonal code is running at four times the bit rate $(1/T_b)$. Therefore, we have an effective bandwidth expansion factor of four. The bandwidth expansion factor is sometimes called the *processing gain* or *(W/R)*, where *W* is the final bandwidth of the spread message and *R* is the bandwidth of the baseband message. Note that in this example, *W* is equivalent to $(1/T_c)$, *R* is equivalent to $(1/T_b)$, and the processing gain *(W/R)* is 4, or 6 dB. For an excellent treatment of power spectra of different digital waveforms, consult [4].

Note that after spreading by the orthogonal codes, the spread messages $m_1(t)c_1(t)$ and $m_2(t)c_2(t)$ now occupy a larger bandwidth than the original messages.

Figure 1.3 shows the waveforms at different points of the receiver. The signal at point A is the result of the summation of the two spread messages. The spectrum at A now contains two separate signals. In order to recover the two separate messages from the composite spectrum, the signal at A is multiplied by the two respective orthogonal codes to obtain B_1 and B_2. Figure 1.4 shows the signals at C_1 and C_2, the outputs $\tilde{m}_1$ and $\tilde{m}_2$ of the decision thresholds, and the recovered messages $\tilde{m}_1(t)$ and $\tilde{m}_2(t)$. The integrator adds up the signal power over one bit interval T_b of the baseband message, and the decision threshold decides, based on the output of the integrator, whether or not the particular bit is a $+1$ or -1. If the output of the integrator is greater than 0, then the decision is a $+1$; if the integrator output is less than 0, then the decision is a -1. The *digital-to-analog* (D/A) converter transforms the decision into the recovered waveforms $\tilde{m}_1(t)$ and $\tilde{m}_2(t)$. As one can see in this idealized example, the recovered messages $\tilde{m}_1(t)$ and $\tilde{m}_2(t)$ match perfectly the original baseband messages $m_1(t)$ and $m_2(t)$.

This example only serves to illustrate the principle of DS-SS multiple access. We have just demonstrated that, using DS-SS techniques, separate messages can be sent through the same channel in the same frequency band at the same time, and the messages can be successfully recovered at the receiver. However, there are many real-world phenomena, especially in a mobile communications environment, that degrade the performance of such a DS-SS multiple

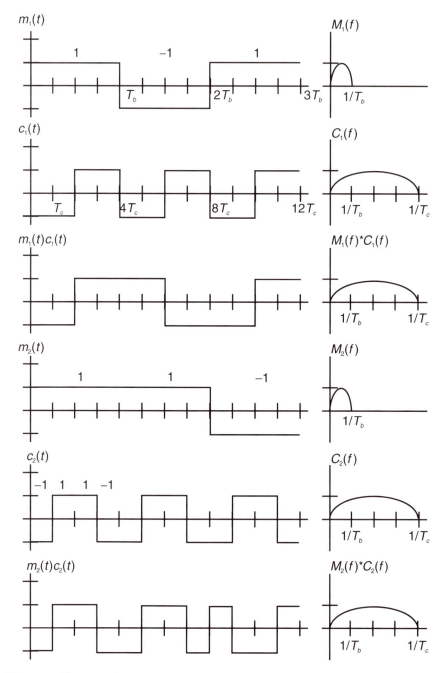

Figure 1.2 Time waveforms and frequency spectra for the baseband messages $m_1(t)$ and $m_2(t)$, orthogonal codes $c_1(t)$ and $c_2(t)$, and spread messages $m_1(t)c_1(t)$ and $m_2(t)c_2(t)$.

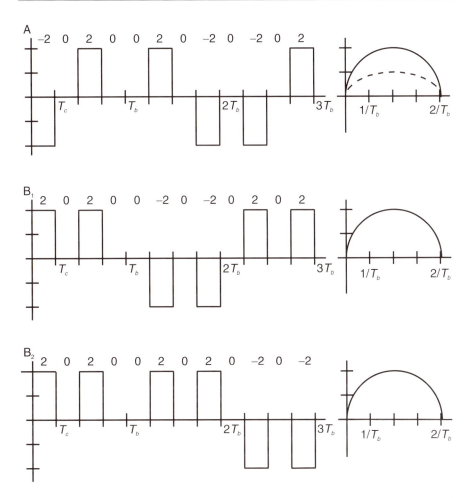

Figure 1.3 Time waveforms and frequency spectra for the signals at different points of the receiver.

access system. There are two problems: the *near-far* problem, and the *partial correlation* problem [2].

In mobile communications, each user is geographically dispersed but transmitting in the same frequency spectrum using DS-SS. Some users are closer to the base station than others. The result is that powers received from those users that are close by are higher than powers received from users that are farther away. Because all users are transmitting in the same band, the higher received powers from users nearby constitute an interference that degrades the general performance of the system. In order to combat this near-far phenomenon, power control is utilized to make sure that the powers received at the base

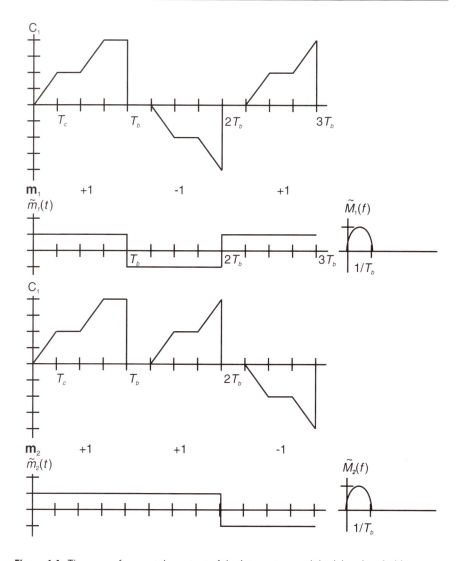

Figure 1.4 Time waveforms at the output of the integrators and decision threshold.

station are the same for all users. In the previous example, we have assumed *perfect* power control by specifying that $m_1(t)c_1(t)$ and $m_2(t)c_2(t)$ both have the same amplitudes (i.e., ranging from +1 to −1). Power control is treated in more detail in Chapter 4 of this book.

The second problem is partial correlation. This problem comes when there is no attempt to synchronize the transmitters sharing the same band. Even when the transmitters are synchronized, there is still the problem of propagation delay, which is inherent in a mobile channel. For example, the two codes mentioned previously are orthogonal when they are perfectly aligned:

x_i	-1	-1	1	1	-1	-1	1	1
y_i	-1	1	1	-1	-1	1	1	-1

However, if y_i suffers a delay of one chip as a result of propagation delay in a mobile channel, then

x_i		-1	-1	1	1	-1	-1	1	1
y_{i-1}	1	1	-1	-1	1	1	-1	-1	

Readers can easily verify that these two sequences are no longer orthogonal. If the codes are not orthogonal due to synchronization or channel impairment, then multiple access messages in the same band can no longer be separated from one another via code-orthogonality. The results are correlation crosstalk and mutual interference. In essence, an additional condition needs to be adhered to; that is,

$$\int_0^{T-\tau} x(t)y(t+\tau)dt = 0$$

$$\int_{T-\tau}^{T} x(t)y(t+\tau-T)dt = 0$$

Therefore, simple orthogonality between two aligned codes is not enough—the above two partial correlations must also be zero, or at least small, for any value of τ likely encountered in the system [2]. Radio propagation is treated in Chapter 2, while techniques to combat partial correlation in CDMA are presented in Chapter 5.

1.3 Applications of DS-SS in Mobile Communication

Despite its difficulties, which are easily solved with optimized system design, CDMA does have its advantages when applied to mobile communications.

First of all, a CDMA system can readily take advantage of the *voice activity* of normal human speech. In a two-person conversation, each speaker is active less than half of the time. During the quiet period, the transmitters could effectively turn off and reduce interference power introduced into the channel. This reduction in interference can translate into capacity gain for the system. Theoretically, FDMA and TDMA systems could also take advantage of the speech statistics. However, the implementation is more complicated as radio resources, such as FDMA channels or TDMA time slots, need to be dynamically assigned in real time by the network infrastructure.

The second advantage is that in CDMA, the physical RF channel can be reused in every cell, thus giving a frequency reuse factor of close to 1. In a conventional AMPS system, the available spectrum is divided into chunks and assigned to different cells in the system. Cochannel frequencies are not used in adjacent cells to avoid interference. A popular frequency-assignment plan is the $N = 7$ reuse pattern, where the spectrum is divided into seven chunks, and each chunk is assigned to one of the cells in a seven-cell cluster. The same chunk is reused again approximately two cells away in the next cluster. The consequence, however, is that the number of channels per cell is reduced by the reuse factor (seven in the $N = 7$ reuse pattern) [5]. The reuse could be increased via sectorization. In CDMA, the same physical channels are used in every cell, but the same cochannel interference problem also exists; on the forward link (i.e., base station to mobile station link), each user in a given cell is being interfered with by powers from its own cell as well as by powers from other cells. On the reverse link (i.e., mobile station to base station link), each cell is being interfered with by users in its coverage area as well as by users located in other cells. There exists no simple analytical solution to quantify the corresponding cochannel interference in CDMA, as the amount of interference depends on the distribution and number of users and terrain. However, there is no need to *frequency plan* in CDMA, which may be one of the welcoming benefits for RF design engineers.

The third advantage is CDMA's ability to mitigate multipath distortion [6]. If multipath distortion is fixed with time, it can be effectively countered by adaptive equalization. If, on the other hand, it is rapidly varying with time, as in a mobile environment, it would be difficult to adapt sufficiently fast. Spread spectrum, and in particular direct-sequence spread spectrum, gives an extra measure of immunity to multipath distortion. This result can be seen clearly in the frequency domain where the multipath distortion leads to a null in the frequency band. This null severely affects a narrowband signal if the null occupies a significant portion of the bandwidth. But the same null would have less effect on a spread broadband signal [2]. Furthermore, a CDMA system can take advantage of multipaths by using the *rake receiver*, which demodulates and uses the signal energy of all paths. The effects of propagation on signal spectrum are discussed in Chapter 2.

References

[1]   Pickholtz, R. L., D. L. Schilling, and L. B. Milstein, "Theory of Spread-Spectrum Communications—A Tutorial," *IEEE Trans. on Communications,* Vol. COM-30, No. 5, May 1982.

[2]   Lee, E. A., and D. G. Messerschmitt, *Digital Communication,* Boston, MA: Kluwer Academic Publishers, 1990.

[3]   Faruque, S., *Cellular Mobile Systems Engineering,* Norwood, MA: Artech House, 1996.

[4]   Carlson, B. A., *Communication Systems,* New York, NY: McGraw-Hill, 1986.

[5]   Viterbi, A. J., *CDMA Principles of Spread Spectrum Communication,* New York, NY: Addison-Wesley, 1995.

[6]   Gilhousen, K. S., et al., "On the Capacity of a Cellular CDMA System," *IEEE Trans. on Vehicular Technology,* Vol. 40, May 1991, pp. 306–307.

Select Bibliography

Glisic, S., and B. Vucetic, *Spread Spectrum CDMA Systems for Wireless Communications,* Norwood, MA: Artech House, 1997.

Harte, L., *CDMA IS-95 for Cellular and PCS: Technology, Applications and Resource Guide,* New York, NY: McGraw-Hill, 1997.

Peterson, R. L., R. E. Ziemer, and D. E. Borth, *Introduction to Spread-Spectrum Communications,* Upper Saddle River, NJ: Prentice Hall, 1995.

Proakis, J. G., *Digital Communications,* New York, NY: McGraw-Hill, 1995.

Wozencraft, J. M., and I. M. Jacobs, *Principles of Communication Engineering,* Waveland Press, 1990.

Yacoub, M. D., *Foundations of Mobile Radio Engineering,* Books Britain, 1993.

2

Radio Propagation

2.1 Link Analysis

In any communication system, we are concerned with one critical parameter, C/N, which is the *carrier-to-noise* ratio at the receiver. This parameter defines how much signal power there is as compared to the noise power over the channel; therefore, C/N can be considered as a figure of merit for the communication system.

The *link equation* is an equation that calculates the C/N using several other parameters of the communication system:

$$\frac{C}{N} = \frac{(\text{ERP})L_p G_r}{N} \qquad (2.1)$$

where ERP is the *effective radiated power* from the transmit antenna, L_p is the propagation loss in the channel, G_r is the gain of the receive antenna, and N is the effective noise power. In particular, ERP is calculated by the following equation:

$$\text{ERP} = P_t L_c G_t \qquad (2.2)$$

where P_t is the power at the output of the transmitter power amplifier, L_c is the cable loss between the power amplifier and transmit antenna, and G_t is the gain of the transmit antenna. Although there are many definitions of effective noise power N, here we constrain our definition of N to just *thermal noise*, which is defined as

$$N = kTW \qquad (2.3)$$

where k is the Boltzmann's constant (1.38×10^{-23}W/Hz/K or -228.6 dBW/Hz/K), T is the noise temperature of the receiver, and W is the bandwidth of the system. In subsequent discussions, we encounter another similar parameter C/I, or *carrier-to-interference* ratio. C/I differs from C/N in that the denominator of C/I includes not only thermal noise power but also interference power from other sources. In mobile communication systems, C/I is a more commonly used figure of merit because it takes other interference effects into account. For now, we use carrier-to-noise ratio as our indicator of link quality.

As one can see from (2.1), the link quality is dependent on parameters such as gains of the transmit and receive antennas, transmitter power, and receiver noise temperature. All these parameters are within the control of the system designer and can be changed to optimize system performance. One parameter, however, in (2.1) is not within the control of the system designer. This parameter is propagation, or path loss. This loss refers to the attenuation the signal suffers en route from the transmitter to the receiver. We discuss in the next section several methods of predicting the propagation loss in a radio environment.

2.2 Propagation Loss

The propagation loss in (2.1) encompasses all the impairments that the signal is expected to suffer as it travels from the transmitter to the receiver. There are many prediction models that are used to predict path loss. Although these models differ in their methodologies, all have the distance between transmitter and receiver as a critical parameter. In other words, the path loss is heavily dependent on the distance between the transmitter and receiver. Other effects may also come into play in addition to distance. For example, in satellite communications, atmospheric effects and rain absorption are dominant in determining received signal power. Here, we describe three models: free space, the Lee model, and the Hata model.

2.2.1 Free-Space Model

In free space, electromagnetic waves diminish as a function of inverse square, or $1/d^2$, where d is the distance between the transmitter and receiver. In its linear form, the free-space path loss is

$$L_p = \frac{4\pi\lambda^2}{d^2}$$

(2.4)

where λ is the wavelength of the signal. Equation (2.4) can also be written in decibel form as shown in the following equation:

$$L_p = -32.4 - 20\log(f) - 20\log(d)$$

(2.5)

where d is in kilometers, f is the frequency of the signal in megahertz, and path loss L_p is in decibels. In deriving (2.5), we use the fact that the speed of light is a product of frequency and wavelength (i.e., $c = \lambda f$). Note that once the carrier frequency of the signal, f, is known, the first and second terms of (2.5) are effectively constants, and L_p varies strictly as a function of d in the third term. If we plot (2.5) on a log-log paper, then the *slope* of the curve would be −20 dB/decade.

The free-space model is based on the concept of an expanding spherical wavefront as the signal radiates from a point source in space. It is mostly used in satellite and deep-space communication systems where the signals truly travel through "free space." In a mobile communication system where additional losses are introduced by terrestrial obstacles and other impairments, alternative models are needed to accurately predict propagation loss.

2.2.2 Lee Model

The propagation environment in terrestrial communication is worse than that in free space. There are often obstacles between the base station and the mobile user. As a result, the received signal is made up of signals traveling via direct and indirect paths. Signals traveling in direct paths are those in *line-of-sight* (LOS), and signals traveling in indirect paths are those involving refraction and reflection from objects (such as buildings, trees, and hills) between the transmitter and the receiver. Therefore, the path loss in a terrestrial environment is higher than that in free space, and the extent of the loss is even more strongly influenced by the distance between the transmitter and the receiver. For illustration purposes, we present a simplified formula of the Lee model at the cellular frequency:

$$L_p = 1.14 \times 10^{-13} \frac{h^2}{d^{3.84}}$$

(2.6)

where d is the distance (in kilometers) between the base station and the mobile user and h is the height (in meters) of the base station antenna. Note that in this case, the path loss varies as an inverse power of 3.84 compared to an inverse power of 2 in free space. In other words, the path loss encountered in terrestrial mobile communication systems is worse than that seen in free space. In addition, the loss becomes less as the base station height h increases (i.e., the loss becomes less severe as the base station antenna is raised higher). Converting (2.6) into decibel form yields

$$L_p = -129.45 - 38.4 \log(d) + 20 \log(h)$$

(2.7)

where, again, d is in kilometers and h is in meters. Note that in (2.7) the path loss slope is -38.4 dB/decade.

The generalized form of the Lee model is much more complicated than that presented in (2.6) and (2.7). The model is quite powerful and contains different parameters to use under various propagation and terrestrial conditions. For a complete treatment of the Lee model, refer to William C. Y. Lee's text [1].

2.2.3 Hata Model

A good propagation model should be a function of different parameters necessary to describe the various propagation conditions. Here, we use the Hata model to illustrate a slightly more complicated path loss model that's a function of parameters such as frequency, frequency range, heights of transmitter and receiver, and building density. The Hata model is based on extensive empirical measurements taken in urban environments. In its decibel form, the generalized model can be written as

$$L_p = -K_1 - K_2 \log(f) + 13.82 \log(h_b) + a(h_m)$$
$$- [44.9 - 6.55 \log(h_b)] \log(d) - K_0$$

(2.8)

where f is the carrier frequency (in megahertz), h_b is the antenna height (in meters) of the base station, h_m is the mobile antenna height (in meters), and d is the distance (in kilometers) between the base station and the mobile user. For these parameters, there are only certain ranges in which the model is valid; that is, h_b should only be between 30m to 200m, h_m should be between 1m to 10m, and d should be between 1 km to 20 km. Note that the slope of (2.8) is $-[44.9 - 6.55\log(h_b)]$ dB/decade.

The terms $a(h_m)$ and K_0 are used to account for whether the propagation takes place in an "urban" or a "dense urban" environment. In particular,

$$a(h_m) = [1.1\log(f) - 0.7]h_m - [1.56\log(f) - 0.8] \text{ for "urban," or}$$
$$a(h_m) = 3.2[\log(11.75h_m)]^2 - 4.97 \text{ for "dense urban"}$$

and

$$K_0 = 0 \text{ for "urban," or}$$
$$K_0 = 3 \text{ dB for "dense urban"}$$

The term K_1 and the factor K_2 are used to account for the frequency ranges. Specifically,

$$K_1 = 69.55 \text{ for frequency range 150 MHz} \le f \le 1000 \text{ MHz, or}$$
$$K_1 = 46.3 \text{ for frequency range 1500 MHz} \le f \le 2000 \text{ MHz}$$

and

$$K_2 = 26.16 \text{ for frequency range 150 MHz} \le f \le 1000 \text{ MHz, or}$$
$$K_2 = 33.9 \text{ for frequency range 1500 MHz} \le f \le 2000 \text{ MHz}$$

Readers are referred to [2] for a complete treatment of the Hata model.

2.2.4 Observations

We note that all the propagation models presented can be written in the general form of a straight-line equation (in decibels):

$$L_p = -L_0 - \gamma \log(d) \tag{2.9}$$

where L_0 is the *intercept* and γ is the *slope*. The slope is a factor showing how severely the signal power decreases as a function of distance. For illustration purposes, Figure 2.1 shows a comparison between the three propagation models: free space, Lee, and Hata. Note that the slope for each of these models is, respectively, −20 dB/decade, −38.4 dB/decade, and −35.2 dB/decade for a base station height of 30m.

These prediction models have their limitations when used to model propagation loss in terrestrial environments. The accuracy of these models typically varies between 6 to 8 dB when compared to field measurements. The accuracy can be increased, however, by integrating the field measurement results with the model. For example, it is a common industry practice to take field measurements and custom calculate the model *slope* that is used over certain distances from the base station.

Another limitation is that the prediction models presented cannot be used over *microcell* regions. The microcell regions refer to those distances that are very close to the base station, typically less than one mile. Other propagation phenomena dominate when one attempts to predict path loss very near the base

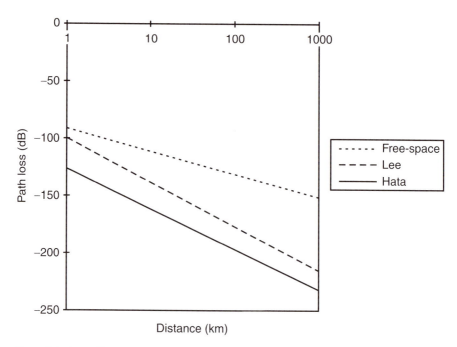

Figure 2.1 As an illustration, the graph shows the path loss vs. distance for three different propagation models: free space, Lee, and Hata. The antenna height and carrier frequency are 30m and 881.5 MHz, respectively. For the Hata prediction, we use a mobile antenna height of 1.5m and an urban scenario.

station; hence, other specialized microcell models are needed to predict losses in these regions. Readers are referred to [3] for a very good description of specialized microcell path loss models.

2.3 Shadowing

The signal power in the direct path decreases relatively slowly as the receiver moves away from the transmitter. However, as a receiver traverses away, obstacles that partially block the signal path (such as trees, building, and moving trucks) cause occasional drops in received power. This decrease in power occurs over many wavelengths of the carrier and is thus called *slow fading*. Slow fading is usually modeled by a *log-normal* distribution with mean power and standard deviation (i.e., the probability distribution of the power variation is distributed as $10^{\xi/10}$, where ξ is a normal, or Gaussian random variable with mean m and standard deviation σ). The standard deviation in a cellular environment is typically around 8 dB. We know that the average received power decreases (due to path loss) as the mobile moves away from the base station. Another way to visualize slow fading is to picture that there is a slow power variation (occurring over many wavelengths) on top of the average, and that variation can be described by a log-normal probability distribution.

The reason for the log-normal distributed slow fading is that the received signal is the result of the transmitted signal passing through or reflecting off many different objects, such as trees and buildings. Each object attenuates the signal to some extent, and the final received signal power is the sum of transmission factors of all these objects. As a consequence, the logarithm of the received signal equates to the sum of a large number of transmission factors, each of which is also expressed in decibels. As the number of factors becomes large, the central limit theorem dictates that the distribution of the sum approaches a Gaussian, even if the individual terms are not Gaussian [4].

2.4 Multipath Rayleigh Fading

There are times when a mobile receiver is completely out of sight of the base station transmitter (i.e., there is no signal path traveling to the receiver via LOS). In this case, the received signals are made up of a group of reflections from objects, and none of the reflected paths is any more dominant than the other ones. The different reflected signal paths arrive at slightly different times, with different amplitudes, and with different phases.

It was verified, both theoretically and experimentally, that the envelope of a received carrier signal for a moving mobile is Rayleigh distributed [5]. Therefore, this type of fading is called *Rayleigh* fading. The theoretical model makes use of the fact that there are many reflected signal paths from different directions (i.e., N signal paths). The composite received signal is

$$r(t) = \sum_{n=1}^{N} R_n \cos(2\pi ft - 2\pi f_{D,n}t)$$

(2.10)

Note that the received signal is made up of N reflected signals; each reflected path has an amplitude of R_n and f is the carrier frequency. The frequency shift $f_{D,n}$ of each reflected signal is due to the Doppler effect when the mobile user is in motion. If the signal is traveling parallel to the mobile's direction of motion, then the Doppler frequency shift is

$$f_{D,n} = \frac{v}{\lambda}$$

(2.11)

where v is the velocity of the mobile. The in-phase and quadrature representation of the received signal in (2.10) is

$$r(t) = R_I(t)\cos(2\pi ft) + R_Q(t)\sin(2\pi ft)$$

(2.12)

where the in-phase component is

$$R_I(t) = \sum_{n=1}^{N} R_n \cos 2\pi f_{D,n}t$$

(2.13)

and the quadrature component is

$$R_Q(t) = \sum_{n=1}^{N} R_n \sin 2\pi f_{D,n}t$$

(2.14)

The terms in the summations of (2.13) and (2.14) are *independent and identically-distributed* (i.i.d.) random variables. Therefore, if N is large, both $R_I(t)$ and $R_Q(t)$ become zero-mean Gaussian random variables. The signal envelope

$$R(t) = \sqrt{R_I^2(t) + R_Q^2(t)} \tag{2.15}$$

then has a Rayleigh distribution. Incidentally, the Rayleigh distribution has the following probability density function:

$$p(R) = \frac{R}{\sigma} e^{\frac{-R^2}{2\sigma^2}} \text{ for } 0 \leq R \tag{2.16}$$

and $p(R) = 0$ for $R < 0$. One way to visualize this type of fading is to picture a base station transmitting an unmodulated carrier with a constant envelope. The received waveform at the mobile would have a varying envelope; the envelope variation is distributed according to a Rayleigh distribution. The bandwidth of this envelope variation is determined by the maximum Doppler frequency shift, which is due to the velocity of the mobile.

Because there are many different signal paths, constructive and destructive interference can result. Thus another way to visualize this particular fading phenomenon is to picture electromagnetic fields radiated by a base station combining constructively and destructively, forming a standing wave pattern in the surrounding area. As a mobile receiver moves through the field, successive drops in amplitudes, or "fades," occur. See Figure 2.2. The distance and spacing between each fade is dependent on the carrier frequency. As a receiver

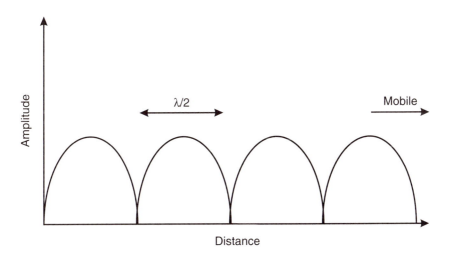

Figure 2.2 For illustration, as it travels through the standing wave pattern, the mobile will experience fades once every half wavelength. Note that the standing wave pattern shown is a simple example resulting from the addition of two equally strong waves that are 180 degrees out of phase.

moves through the field, the rate of change of received amplitude and phase is thus dependent on both the carrier frequency and the receiver velocity. In a mobile environment, the amplitude variation due to this fading phenomenon could be on the order of 50 dB. Because this type of fading could occur very rapidly, it is sometimes called *fast fading*.

Example 2.1

Compare the rate of fade occurrence between cellular and PCS services. Assume that the mobile user is traveling at 90 km/hr, or 25 m/sec.

As an approximation, we use 900 MHz as the cellular carrier frequency and 1.9 GHz as the PCS carrier frequency. The wavelengths are

$$\lambda_{\text{cellular}} = \frac{c}{f_{\text{cellular}}} = \frac{3 \times 10^8}{900 \times 10^6} = 0.33\text{m}$$

$$\lambda_{\text{PCS}} = \frac{c}{f_{\text{PCS}}} = \frac{3 \times 10^8}{1.9 \times 10^9} = 0.16\text{m}$$

The time it takes a mobile user to travel from one fade to the next fade is

$$\Delta t_{\text{cellular}} = \frac{(\lambda_{\text{cellular}} / 2)}{v} = \frac{0.167\text{m}}{25\text{m/sec}} = 6.67\text{m sec}$$

$$\Delta t_{\text{PCS}} = \frac{(\lambda_{\text{PCS}} / 2)}{v} = \frac{0.079\text{m}}{25\text{m/sec}} = 3.16\text{m sec}$$

Therefore, for cellular, we can expect to see a significant drop in signal strength, or fade, once every 6.67 msec, or at a rate of 150 Hz. For PCS, we experience one fade every 3.16 msec, or at a rate of 317 Hz. Incidentally, the Doppler frequency shifts for these two cases are

$$f_{\text{D, cellular}} = \frac{v}{\lambda_{\text{cellular}}} = \frac{25\text{m/sec}}{0.33\text{m}} = 75\text{Hz}$$

$$f_{D,PCS} = \frac{v}{\lambda_{PCS}} = \frac{25\text{m/sec}}{0.16\text{m}} = 158\text{Hz}$$

2.5 Multipath Delay Spread

Multipath occurs when signals arrive at the receiver directly from the transmitter and, indirectly, due to transmission through objects or reflection. The amount of signal reflection depends on factors such as angle of arrival, carrier frequency, and polarization of incident wave. Because the path lengths are different between the direct path and the reflected path(s), different signal paths could arrive at the receiver at different times over different distances. Figure 2.3 illustrates the concept. An impulse is transmitted at time 0; assuming that there are a multitude of reflected paths present, a receiver approximately 1 km away should detect a series of pulses, or *delay spread*.

If the time difference Δt is significant compared to one symbol period, *intersymbol interference* (ISI) can occur. In other words, symbols arriving significantly earlier or later than their own symbol periods can corrupt preceding or trailing symbols. For a fixed-path difference and a given delay spread, a higher data rate system is more likely to suffer ISI due to delay spread. For a fixed data

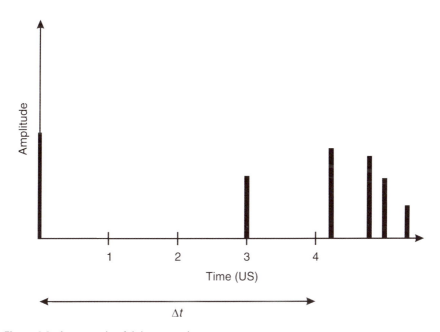

Figure 2.3 An example of delay spread.

rate system, a propagation environment with longer path differences (and thus higher delay spread) is more likely to cause ISI.

Example 2.2

Determine if the delay spread profile shown in Figure 2.3 will cause ISI to a mobile communication system using 270.83 Kbps as its data rate:

$$R_b = 270.83 \, \text{Kbps}$$

$$T_b = \frac{1}{R_b} = \frac{1}{270.83 \times 10^3 \, \text{bps}} = 3.69 \, \mu \sec$$

Since the bit period is approximately the same as the delay shown in Figure 2.3, ISI could occur in this situation without any use of equalization.

Example 2.3

Determine if the delay spread profile shown in Figure 2.3 will cause ISI to a mobile communication system using 1.2288 Mbps as its bit rate (as in IS-95 CDMA):

$$R_b = 1.2288 \, \text{Mbps}$$

$$T_b = \frac{1}{R_b} = \frac{1}{1.2288 \times 10^6 \, \text{bps}} = 0.81 \, \mu \sec \approx 1 \, \mu \sec$$

Since in this case the delay is so much more than the bit period, ISI would normally occur. However, an IS-95 CDMA system uses a special form of time diversity to recover the signal. The system uses a *rake receiver* to lock onto the different multipath components. If a time reference is provided, then the different multipath components can be separately identified as distinct echoes of the signal separated in time. These separately identified components of the received signal can then be brought in phase and combined to yield a final composite received signal [6]. However, the IS-95 CDMA system cannot separately identify, or resolve, multipath components that are less than 1 μsec apart. In a dense urban environment such as New York, where base stations are very close to each other and each base station is operating at low power, multipath components may arrive at intervals less than 1 μsec with very small power. In this case, IS-95 CDMA would not be able to resolve the components and combine their powers to yield a usable signal. This is one of the reasons a new variant of

CDMA, called *broadband CDMA* (B-CDMA), has been proposed. The B-CDMA variant has a bit rate of 5 Mbps, and it can theoretically resolve multipath components that are $0.2\,\mu$sec apart.

We can also examine the effect of delay spread in the frequency domain. *Delay spread* in the time domain translates directly into *frequency-selective fading* in the frequency domain. Let's use a simple model to illustrate. We assume there are two multipaths having the same amplitude A, as shown in Figure 2.4. One multipath is delayed by τ relative to the other multipath. The received signal is

$$r(t) = As(t) + As(t - \tau) \tag{2.17}$$

By taking the Fourier transform, we arrive at the spectrum of $r(t)$:

$$R(f) = AS(f) + AS(f)e^{-j2\pi f \tau}$$

which can be rewritten as

$$R(f) = AS(f)\left[1 + e^{-j2\pi f \tau}\right] = AS(f)H(f)$$

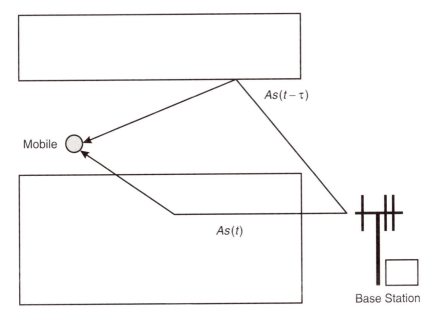

Figure 2.4 Two multipath components separated by time τ.

Here, $H(f)$ is effectively the transfer function of the channel that transforms the original signal $AS(f)$. $H(f)$ can also be written as

$$H(f) = 1 + e^{-j2\pi f\tau} = e^{-j2\pi f(\tau/2)}\left[e^{j2\pi f(\tau/2)} + e^{-j2\pi f(\tau/2)}\right]$$
$$= 2e^{-j2\pi f(\tau/2)}\cos(2\pi f(\tau/2))$$

and the magnitude spectrum $|H(f)|$ is

$$|H(f)| = 2\cos(2\pi f(\tau/2)) \tag{2.18}$$

$|H(f)|$ is shown in Figure 2.5. The frequency-selective fading is thus evident in the nulls of the magnitude spectrum as a result of multipath delay.

Exercise 2.1

Verify that the B-CDMA signal (5 Mbps) is more advantageous than the IS-95 CDMA signal (1.2288 Mbps) in terms of frequency-selective fading if $\tau = 0.4\,\mu\text{sec}$.

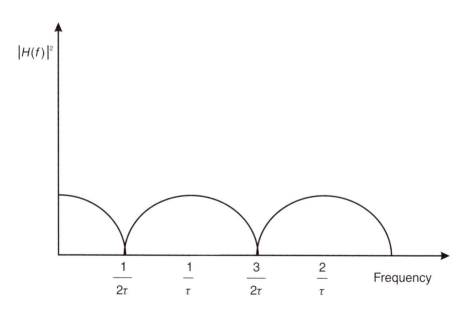

Figure 2.5 Frequency-selective fading is evident in the nulls of the transfer function.

References

[1]   Lee, W. C. Y., *Mobile Cellular Telecommunications: Analog and Digital Systems*, New York, NY: McGraw-Hill, 1995.

[2]   Hata, M., "Empirical Formula for Propagation Loss in Land Mobile Radio Services," *IEEE Trans. on Vehicular Technology*, Vol. VT-29, 1980, pp. 317–325.

[3]   Lee, W. C. Y., *Mobile Communications Design Fundamentals*, New York, NY: John Wiley & Sons, 1993.

[4]   Hess, G. C., *Land-Mobile Radio System Engineering*, Norwood, MA: Artech House, 1993.

[5]   Jakes, W. C., *Microwave Mobile Communications*, New York, NY: John Wiley & Sons, 1974.

[6]   Mehrotra, A., *Cellular Radio Performance Engineering*, Norwood, MA: Artech House, 1994.

Select Bibliography

Lee, W. C. Y., *Mobile Communications Engineering*, New York, NY: McGraw-Hill, 1997.

Lee, W. C. Y., "Lee's Model," *Proc. 42nd Annual Vehicular Technology Conf.*, IEEE, 1992, pp. 343–348.

Rappaport, T. S., *Wireless Communications: Principles and Practice*, Englewood Cliffs, NJ: Prentice Hall, 1995.

Yacoub, M. D., *Foundations of Mobile Radio Engineering*, Books Britain, 1993.

3

Fundamentals of Digital RF Communication

3.1 Introduction

In this chapter, we address some of the fundamental issues in digital communication, particularly as applied to RF applications such as digital mobile radio. We approach the subject from a *system* perspective. Namely, we treat the subject in terms of subsystems and block diagrams that carry out specific functions in the overall scheme of moving information from the transmitter to the receiver. Since this book deals primarily with mobile radio systems using DS-SS techniques, we tailor our discussion to digital communication systems that are used by the IS-95 CDMA standard.

Before we begin, it is important to understand why the wireless personal communications industry is moving from the traditional analog FM technology to digital technologies such as TDMA and CDMA. There are many advantages for migrating to digital, but there are at least four reasons for the recent trend in the industry, as follows.

The first is the *quality of service*. Digital communication systems, as manifested in various technologies, have the potential to offer a higher quality of service. This is particularly true in harsh radio environments such as the mobile channel. This improved service is made possible by a digital system's inherent ability to regenerate the signal. Figure 3.1 illustrates the point. In this simplified digital communication system, a positive pulse (which could be designated as a 1) is sent from the transmitter. As the pulse travels to the receiver, it suffers impairment in both *amplitude* and *shape*. The amplitude of the pulse is due to

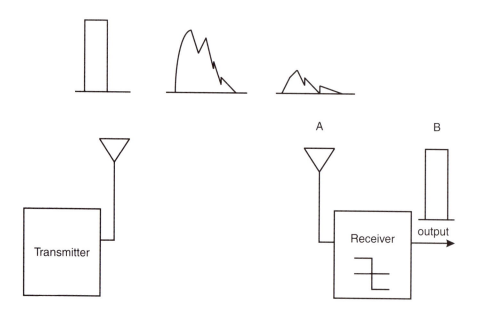

Figure 3.1 A simplified digital communication system showing the concept of signal regeneration.

propagation loss, which is typically a function of distance. The shape of the pulse is also distorted, partly due to the lowpass nature of the channel.

At point A, the signal has suffered so much degradation that if it were an analog signal, the low signal-to-noise ratio would result in poor signal quality. However, the digital communication system shown has a receiver that contains a *threshold detector*. The detector is programmed to output a 1 if the cumulative energy over the specified bit period is greater than 0, and to output a −1 if the cumulative energy over the period is less than 0. In this case, the cumulative energy is greater than 0, thus the threshold detector makes a *decision* of 1 and perfectly regenerates the transmitted pulse.

In this example, the information is contained in the *amplitude* of the signal. Other digital communication systems, such as the CDMA system, use *phase-shift keying* (PSK), where the information is contained in the *phase* of the signal. The characteristics of PSK are discussed in the modulation section later in this chapter.

The second reason is *increased capacity*. Because a digital communication system transmits its information in discrete units, namely 1 and −1 (or 1 and 0), source information must be converted to discrete units as well. In a voice communication system, the source information is human speech; speech is inherently analog and must be converted (coded) to a digital form before it can be transmitted via the digital system. This coding of source information, or *source coding*, can use speech compression to reduce the number of bits necessary to represent speech. For example, the *advanced mobile phone system* (AMPS) transmits speech using a 30-kHz bandwidth. The IS-95 CDMA system uses a *vocoder*, or a voice coder, to convert human speech to 9.6 Kbps (Rate Set 1) of digital stream. The vocoder therefore uses a smaller bandwidth to transmit speech information. Moreover, the vocoder employed in CDMA is *variable* rate, which means that during silent period of human speech, the output bit rate of the vocoder is lowered. The IS-95 vocoder supports four different rates: 9,600, 4,800, 2,400, and 1,200 bps. The variable-rate feature further reduces the effective bandwidth required to transmit speech. If there is only a fixed amount of total bandwidth, a reduction in the bandwidths of individual voice channels implies an increase in the total number of available voice channels.

The third reason is *privacy*. Privacy is an important issue in any communication system. In the conventional analog FM system, anyone who has an FM scanner can eavesdrop on a conversation. A digital communication system provides a ready platform where *encryption* techniques may be used to safeguard the information transmitted over the air. In a complex system such as CDMA, it would be difficult for anyone to build a receiver and eavesdrop on a conversation, even if the encryption feature is not activated.

Perhaps the best reason for the industry's recent migration is *enabling technology and economics*. Complex digital communication systems, especially spread-spectrum ones, are made possible with enabling technologies such as *digital signal processing* (DSP) techniques and their implementation on *application-specific integrated circuits* (ASICs). The cost of these technologies has historically been high enough that they could only be afforded by government and military customers. In recent years, however, the economics of these technologies has made them feasible for use in commercial application on a mass scale.

3.2 System Components

Figure 3.2 shows a functional block diagram of a typical digital communication system [1]. The information source, such as human speech, is first converted into digital form by the *source encode* function. Then the *channel encode* function encodes the digital information for the purpose of combating various degrading effects of the channel. Then the information is arranged by the *multiple access* function so that more than one user can share the given spectrum. The *modulate* function converts the information from baseband to a bandpass

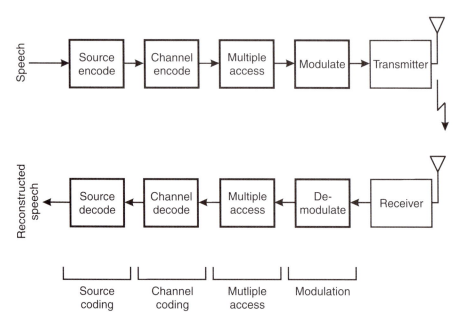

Figure 3.2 Principal components of a digital communication system.

(RF) waveform that can be transmitted by the transmitter. On the receive side, the bandpass waveform is intercepted by the receiver. The signals are first *demodulated* from RF to baseband, then the *multiple access* function separates the different users that are sharing the particular spectrum. Then, the *channel decode* function attempts to correct the errors that have been introduced by the channel. The *source decode* function converts the baseband information back to analog speech.

3.3 Source Coding

The source information has to be coded into a digital form in order for it to be further processed by the digital communication system. One of the techniques used in wireline applications is *pulse code modulation* (PCM), where the analog voice is converted into a 64-Kbps bit stream. Other wireline techniques, such as *adaptive pulse code modulation* (ADPCM) and *delta modulation* (DM), are also used. These source coding schemes for speech use what is called "waveform coding," where the goal is to replicate the waveform of the source information. This is the reason why computer modems can be used over telephones; the information contained in the waveform generated by a transmitting modem can be reliably received by the receiving modem on the other end, and the reason is that PCM attempts to replicate the waveform regardless of whether or not the information contained in the waveform is human speech or modulated pitches generated by a modem.

PCM is not feasible in wireless applications because there is a limited bandwidth available. Transmitting 64 Kbps of information over the air demands more bandwidth than can be afforded by most service providers. Therefore, alternative source coding techniques are needed to represent source information (human speech, in this case) using less bandwidth. A vocoder offers an attractive solution. It exploits the characteristics of human speech and uses fewer bits to represent and replicate human sounds. See Figure 3.3.

3.3.1 Characteristics of Human Speech

Before we discuss vocoding, it is important that we gain a basic understanding of human speech. The temporal and frequency characteristics of human sound are exploited by vocoders for speech coding. The human voice is made up of a combination of *voiced* and *unvoiced* sounds. The voiced sounds such as vowels ("eee" and "uuu") are produced by passing quasi-periodic pulses of air through the vocal tract. These sounds have essentially a periodic rate with a fundamental

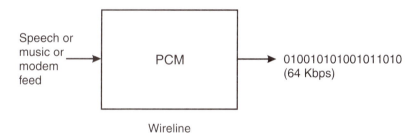

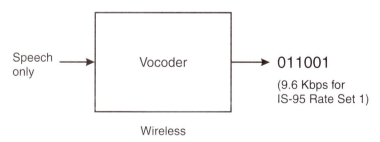

Figure 3.3 Motivation for using vocoders in wireless applications.

frequency. This fundamental frequency is also known as pitch. The unvoiced sounds, such as consonants ("t" and "p"), are produced by passing turbulent air through the vocal tract. These sounds are more like acoustic noise created by a closure and sudden release of vocal tract. Figure 3.4 illustrates the principle of sound generation.

Although human voice is time varying, its spectrum is typically stationary over a period between 20 and 40 ms. This is the reason why most vocoders produce frames that have a duration on this order. For example, the IS-95 vocoder produces frames that are 20 ms in duration.

3.3.2 Vocoders

The voice tract can be modeled by a linear filter that is time varying. That is, the filter response varies with time. This is done by periodically updating the coefficients of the filter. This filter is typically all-pole because an all-pole filter requires less computational power than a filter with both poles and zeros. Thus,

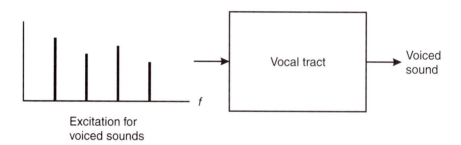

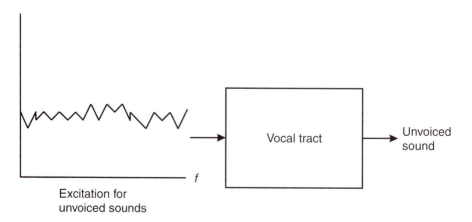

Figure 3.4 The excitation and generation of human voice.

the filter modeling the vocal tract can be represented as $1/T(z)$. If we represent the excitation signal as $E(z)$, then the spectrum of the speech signal $S(z)$ can be written as

$$S(z) = \frac{E(z)}{T(z)} \qquad (3.1)$$

The all-pole filter $1/T(z)$ can be written as

$$\frac{1}{T(z)} = \frac{1}{1 - \sum_{k=1}^{K} b_k z^{-k}} \qquad (3.2)$$

Equation (3.1) can also be written as

$$E(z) = S(z)T(z) \qquad (3.3)$$

The all-zero filter $T(z)$ is sometimes referred to as the analysis filter, and (3.3) represents the process of speech *analysis*. The all-pole filter $1/T(z)$ is referred to as the synthesis filter; it is used in conjunction with the excitation signal $E(z)$ to synthesize the speech signal $S(z)$. Equation (3.1) thus represents the process of speech *synthesis*. This type of coding technique is sometimes called analysis-synthesis coding. Figure 3.5 shows how speech is analyzed at the transmitting end and synthesized at the receiving end. The voice encoder analyzes the speech and produces excitation parameters (such as voiced/unvoiced excitation decisions) and filter coefficients valid over the 20-ms interval. The excitation parameters and filter coefficients are the outputs of the speech encoder. In the IS-95 CDMA system, these parameters and coefficients are the information that is communicated between the transmitter and receiver. The voice decoder at the receiving end uses these parameters and coefficients to construct the excitation source and synthesis filter. The result is estimated speech $\tilde{S}(z)$ at the output of the voice decoder.

Linear-predictive coding (LPC) is widely used to estimate filter coefficients. A feedback loop in the encoder is used to compare actual voice and replicated voice. The difference between actual voice and replicated voice is the *error*. LPC is set up to generate filter coefficients such that this error is minimized. These filter coefficients, along with excitation parameters, are then used by the decoder for speech synthesis.

The IS-95 CDMA system uses a variant of the LPC called *code-excited linear prediction* (CELP). Instead of using the voiced/unvoiced decision, CELP

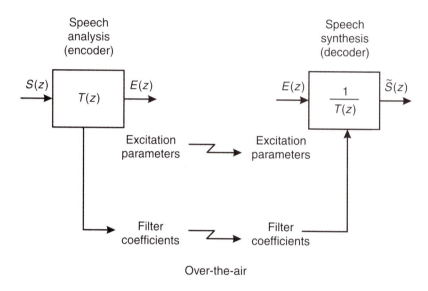

Figure 3.5 Process of replicating human speech.

has a different form of excitation for the all-pole filter. Specifically, the CELP decoder uses a codebook to generate excitation inputs to the synthesis filter. For a complete description of CELP, see Schroeder and Atal [2].

3.4 Channel Coding

After the source information is coded into a digital form, *redundancy* needs to be added to this digital baseband signal. This is done to improve performance of the communication system by enabling the signal to better withstand the effects of channel impairments, such as noise and fading. The goal of channel coding is, given a desired probability of error, to reduce the required E_b/N_0, or alternatively, given an achievable E_b/N_0, to reduce the probability of error. The cost of this goal is more bandwidth, or more redundant bits that the system has to transmit [1].

In this section, we deal specifically with *error-correcting codes*, which when applied to channel coding improve the error performance of the system. The purpose is to add extra bits to the information bits so that errors may be found and corrected at the receiver. In other words, a sequence of bits is represented by a longer sequence of bits with enough redundancy to protect the data [3]. For example, the simplest error-correcting code is to repeat the information bits. Suppose I have a bit I wish to send and error protect. I can simply repeat the bit three times (i.e., if I have 1, I will send 111). This way, I will improve the chance that the receiver correctly receives a 1 in case any one of the transmitted bits is flipped to 0 during the transmission process. In this case, the receiver will use *majority decoding*. Namely, the receiver will only decide a 1 if a majority of the three bits are received as 1s. This code is known as a (3, 1) code. (n, k) refers to a code where k is the length of the information sequence and n is the length of the coded sequence. A code is sometimes described by its rate. The rate R of a code is defined as

$$R = \frac{k}{n}$$

There are two major classes of error-correcting codes: block codes and convolutional codes. Block codes, as the name implies, code an information sequence one block at a time. Convolutional codes, on the other hand, have a memory property. The memory depends on the constraint length K of the convolutional code. The n-tuple output of a convolutional encoder is not only a function of one input k-tuple, but also a function of the previous $K-1$ input k-tuples [1].

3.4.1 Linear Block Codes

3.4.1.1 Minimum Distance

Linear block code is a class of codes that can be used for the purpose of error detection or error correction. A linear block code can be characterized by the (n, k) notation, and for a given code, the encoder transforms a block of k information bits into a longer block of n code bits [4]. The code bits are only a function of the current block of information bits. For example, we can define a (7, 4) linear block code where a block of seven code bits is used to represent a block of four information bits. Given the four information bits (i_1, i_2, i_3, i_4), the three extra *redundancy* bits (r_1, r_2, r_3) are appended using the following functions [3]:

$$r_1 = i_1 + i_2 + i_3$$
$$r_2 = i_2 + i_3 + i_4$$
$$r_3 = i_1 + i_2 + i_4$$

where + represents modulo-2 addition. For example, if the information bits are (1, 0, 1, 0) corresponding to (i_1, i_2, i_3, i_4), then the extra redundancy bits are

$$r_1 = 1 + 0 + 1 = 0$$
$$r_2 = 0 + 1 + 0 = 1$$
$$r_3 = 1 + 0 + 0 = 1$$

and the code word (1, 0, 1, 0, 0, 1, 1) is used to represent the four information bits. Table 3.1 is a complete enumeration of this (7, 4) linear block code.

This simple (7, 4) linear block code is also known as the (7, 4) *Hamming code*, and the redundancy bits are also known as the *parity* bits.

It is intuitive that the extra redundancy bits improve the error performance of the system. To quantify this performance, we introduce the concept of *Hamming distance*. The Hamming distance between any two code words is the number of places that the two code words differ. For example, the Hamming distance between (1, 1, 1, 1, 1, 1, 1) and (1, 1, 1, 0, 1, 0, 0) is 3.

The *minimum distance* d^* of a code is the Hamming distance of a pair of code words with the smallest Hamming distance. For the Hamming code shown above, d^* is 3, which is the smallest Hamming distance for all possible pairs of code words. Minimum distance turns out to be a critical parameter that

Table 3.1
(7,4) Hamming Code

Information Sequence	Redundancy Bits	Code Sequence
0 0 0 0	0 0 0	0 0 0 0 0 0 0
0 0 0 1	0 1 1	0 0 0 1 0 1 1
0 0 1 0	1 1 0	0 0 1 0 1 1 0
0 0 1 1	1 0 1	0 0 1 1 1 0 1
0 1 0 0	1 1 1	0 1 0 0 1 1 1
0 1 0 1	1 0 0	0 1 0 1 1 0 0
0 1 1 0	0 0 1	0 1 1 0 0 0 1
0 1 1 1	0 1 0	0 1 1 1 0 1 0
1 0 0 0	1 0 1	1 0 0 0 1 0 1
1 0 0 1	1 1 0	1 0 0 1 1 1 0
1 0 1 0	0 1 1	1 0 1 0 0 1 1
1 0 1 1	0 0 0	1 0 1 1 0 0 0
1 1 0 0	0 1 0	1 1 0 0 0 1 0
1 1 0 1	0 0 1	1 1 0 1 0 0 1
1 1 1 0	1 0 0	1 1 1 0 1 0 0
1 1 1 1	1 1 1	1 1 1 1 1 1 1

specifies the performance of a particular code. If t errors occur during the transmission of a code word, and if the (Hamming) distance between the received word and every other code word is larger than t, then the decoder will properly correct the errors if it assumes that the closest code word to the received word was actually transmitted [3]. In other words,

$$d^* \geq 2t + 1 \qquad (3.4)$$

If (3.4) holds for a code, then this code is capable of *correcting* t errors. On the other hand, equation (3.5) summarizes the error detection capability q of a code.

$$d^* \geq q + 1 \qquad (3.5)$$

If (3.5) holds for a code, then the code is capable of *detecting* q errors. Thus, given that d^* of the (7, 4) Hamming code is 3, the (7, 4) Hamming code is capable of correcting $t = 1$ error and detecting $q = 2$ errors.

As mentioned above, in order to decode a received code word, the decoder assumes that the closest code word to the received code word was actually transmitted. For example, suppose that a received code word is (0, 0, 0, 1, 1, 1, 1); since this received code word is not one of the specified code words in the (7, 4) Hamming code, an error (or errors) must have occurred. Assuming that the closest code word to the received code word was actually transmitted, the decoder decides that the code word (0, 0, 0, 1, 0, 1, 1) was actually sent by the transmitter. In actuality, a digital logic circuit is used to implement the decoder.

3.4.1.2 Cyclic Redundancy Check (CRC)

IS-95 CDMA uses block coding to indicate the quality of each transmitted frame (which contains a block of information bits). The IS-95 system uses *cyclic redundancy check* (CRC), which is one of the most common block codes. For CRC, the information bits are treated as one long binary number. This number is divided by a unique *prime* number that is also binary, and the remainder is appended to the information bits as redundancy bits. When the frame is received, the receiver performs the same division using the same prime divisor and compares the calculated remainder with the remainder received in the frame [5].

For example, the (7, 4) Hamming code discussed in Section 3.4.1.1 can be generated using a prime divisor of (1, 0, 1, 1). The method can be more clearly seen if we represent binary bits (or a binary number) in a *polynomial* form. For example, the binary bits or number (1, 0, 1, 1) can be represented as a polynomial:

$$g(x) = x^3 + x + 1$$

where each term in the polynomial corresponds to each on-bit of the binary number. The polynomial $g(x)$ is a *prime polynomial.*

Suppose the message (1, 0, 1, 0) needs to be encoded using the (7, 4) Hamming code. To do so, we first convert the message into its polynomial form; that is,

$$m(x) = x^3 + x$$

Then, we shift the message up by $(n - k)$ positions. This can be done very easily in the polynomial form by multiplying the message polynomial $m(x)$ by x^{n-k}. In this case, $(n - k) = (7 - 4) = 3$, so we multiply $m(x)$ by x^3:

$$x^3 m(x) = x^6 + x^4$$

Note that this polynomial corresponds to (1, 0, 1, 0, 0, 0, 0).

The redundancy bits can be obtained by dividing $x^3 m(x)$ by $g(x)$, or

$$x^6 + x^4 = (x^3 + 1)(x^3 + x + 1) + (x + 1)$$

where $(x^6 + x^4)$ is $x^3 m(x)$, $(x^3 + 1)$ is the quotient, $(x^3 + x + 1)$ is the generator polynomial $g(x)$, and $(x + 1)$ is the remainder. The remainder polynomial $(x + 1)$ represents the redundancy bits to be appended to the message; that is, the redundancy bits are (0, 1, 1). As we can see in the (7, 4) Hamming code in Table 3.1, (0, 1, 1) are indeed the redundancy bits to be appended to the message (1, 0, 1, 0). See [6] for a good discussion on cyclic redundancy codes. References [3] and [7] give a good discussion on cyclic codes in general.

In the IS-95 CDMA system, when the vocoder is operating at full rate, each 20-ms frame contains 192 bits, which are made up of 172 information bits, 12 frame quality bits, and 8 encoder tail bits. The 8 encoder tail bits are all set to 0. The 12 frame quality indicator bits are the redundancy bits that are a function of the 172 information bits in the frame. Incidentally, the generator polynomial used to generate the redundancy bits for a full-rate frame is

$$g(x) = x^{12} + x^{11} + x^{10} + x^9 + x^8 + x^4 + x + 1$$

At half-rate, each 20-ms frame contains 96 bits, which are made up of 80 information bits, 8 frame quality bits, and 8 encoder tail bits. In this case, the 8 frame quality bits (or redundancy bits) are generated using the following polynomial:

$$g(x) = x^8 + x^7 + x^4 + x^3 + x + 1$$

These generator polynomials are applied to each frame individually [8].

3.4.2 Convolutional Codes

The block codes are said to be *memoryless*, which means that the code word or the additional CRC bits (in the case of IS-95) are only a function of the current block. The convolutional codes, on the other hand, do have memory. In addition to using CRC, IS-95 CDMA uses convolutional coding to further improve the error performance. For convolutional codes, the encoded bits are functions of information bits and functions of the constraint length. Specifically, every encoded bit (at the output of the convolutional encoder) is a linear combination of some previous information bits. The forward link (base station to mobile station) uses a rate 1/2 and a constraint length $K = 9$ convolutional code. Figure 3.6 shows the convolutional coding scheme for the forward link [8].

Initially, all the registers are initialized to zero. As the information message bits m_i are clocked in from the left, bits are tapped off different stages of the delay line and summed in the modulo-2 adder. The summation is the output of the convolutional encoder. Note that since this is a rate 1/2 coder, two bits are generated for each clock cycle. A commutator switch toggles through both output points for every input clock cycle; hence, the output rate is

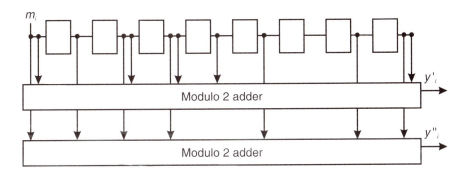

Figure 3.6 Convolutional coding in the IS-95 CDMA system (forward link).

effectively twice the input rate. The generator function for the two output bits y'_i and y''_i (shown in Figure 3.6) can also be written as

$$g'(x) = x^8 + x^7 + x^5 + x^3 + x^2 + x + 1$$
$$g''(x) = x^8 + x^5 + x^4 + x^3 + x^2 + 1$$

IS-95 CDMA uses a different convolutional coding scheme on the reverse link (mobile station to base station). Since the mobile has a limited transmit power, the reverse link can sometimes be the limiting link. Thus a more powerful convolutional code of rate 1/3 and constraint length $K = 9$, is used. In this case, three bits are generated for every input bit, and the output rate is effectively three times the input rate. Figure 3.7 shows the convolutional coding scheme for the reverse link [8].

For reference, the generator function for the three output bits y'_i, y''_i, and y'''_i are

$$g'(x) = x^8 + x^7 + x^6 + x^5 + x^3 + x^2 + 1$$
$$g''(x) = x^8 + x^7 + x^4 + x^3 + x + 1$$
$$g'''(x) = x^8 + x^5 + x^2 + x + 1$$

The decoding mechanism for convolutional codes is beyond the scope of this book. It suffices to mention that convolutional decoding uses a tree search algorithm through a "trellis." The algorithm is a variant of *linear dynamic programming*. See [7] for a good discussion of convolutional decoding.

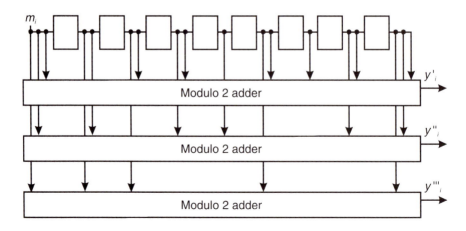

Figure 3.7 Convolutional coding in the IS-95 CDMA system (reverse link).

3.4.3 Interleaving

Signals traveling through a mobile communication channel are susceptible to fading (see Chapter 2). The error-correcting codes are designed to combat errors resulting from fades and, at the same time, keep the signal power at a reasonable level. Most error-correcting codes perform well in correcting *random* errors. However, during periods of deep fades, long streams of successive or *burst* errors may render the error-correcting function useless.

Interleaving is a technique for randomizing the bits in a message stream so that burst errors introduced by the channel can be converted to random errors. In Figure 3.8, we want to send the message "ARE YOU SURE THAT THEY ARE COMING TO LUNCH WITH US" over a fading channel. One way to interleave the message is to load it into a matrix of four rows and ten columns. We truncate the message into four parts and load them into the four rows. Then we read the message out from the top, column by column. The resulting randomized message is sent through the channel.

The channel introduces several burst errors into the message. As a result, the underlined alphabets are received in error. At the receiving end, a *deinterleaver* reconstructs the message using the same matrix, except in this case the deinterleaver loads the received message into columns first, then reads the message out from the rows. As we can see, the burst errors are indeed converted to scattered random errors. In this case, the interleave depth is 10.

The IS-95 CDMA system uses interleaving for the same purpose. The interleaver for the reverse traffic channel uses a matrix of 32 rows by 18 columns (at full rate). The interleaver for the forward traffic channel uses a matrix of 24 rows by 16 columns (at full rate).

3.5 Multiple Access

After the baseband signal has been channel coded for error control, the signal is further transformed in order to allow multiple access by different users. Multiple access refers to the sharing of a common resource in order to allow simultaneous communications by multiple users, and this common resource is the RF spectrum.

In the traditional FDMA scheme, each individual user is assigned a particular frequency band in which transmission can be carried out (see Figure 3.9). A portion of the frequency spectrum is divided into different channels. Different users' signals are lowpass filtered and modulated onto an assigned carrier frequency f_c of a particular channel. This way, multiple users can simultaneously share the frequency spectrum. In TDMA, each user is

Original message:

ARE YOU SURE THAT THEY ARE COMING TO LUNCH WITH US

Interleave matrix:

A   R   E   Y   O   U   S   U   R   E

T   H   A   T   T   H   E   Y   A   R

E   C   O   M   I   N   G   T   O   L

U   N   C   H   W   I   T   H   U   S

Interleaved message:

ATEU RHCN EAOC YTMH OTIW UHNI SEGT UYTH RAOU ERLS

Interleaved message (with burst errors):

ATEU RHCN EAOC YTMH OTI W UHNI SEGT UYT H RAOU ERLS

Reconstructed message (with random errors):

ARE YOU SURE THAT THEY ARE COMING TO LUNCH WITH US

Figure 3.8 An example of interleaving.

assigned a different time slot in which to transmit; in this case, the division of users occurs in the time domain.

In CDMA, each user's narrowband signal is spread over a wider bandwidth. This wider bandwidth is greater than the minimum bandwidth required

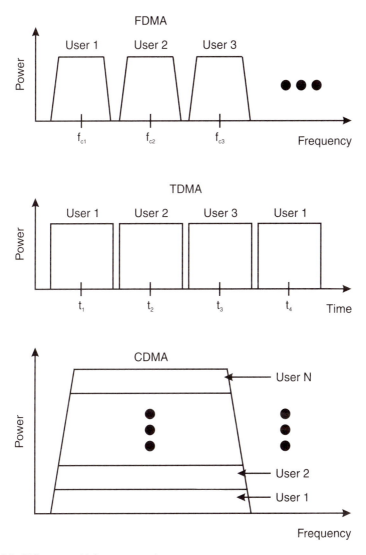

Figure 3.9 Different multiple-access schemes.

to transmit the information. Each user's narrowband signal is spread by a different wideband code. Each of the codes are orthogonal to one another, and channelization of simultaneous users is achieved by the use of this set of orthogonal codes. All the spread wideband signals (of different users) are added together to form a composite signal, and the composite signal is transmitted over the air in the same frequency band. The receiver is able to distinguish among the different users by using a copy of the original code. The receiver sifts the desired user out of the composite signal by correlating the composite signal with the original code. All other users with codes that do not match the code of the desired user are rejected.

The IS-95 CDMA system has *asymmetric links* (i.e., the forward and the reverse links have different link structures). The differences range from the modulation scheme to error control methods. In addition, each link uses different codes to channelize individual users. The forward link uses *Walsh codes*, while the reverse link uses *pseudorandom noice* (PN) *codes* for channelization.

3.5.1 Walsh Codes

3.5.1.1 Generation of Walsh Codes

Figure 3.9 shows that in a CDMA system, all the users are transmitted in the same RF band. In order to avoid mutual interference on the forward link, Walsh codes are used to separate individual users while they simultaneously occupy the same RF band. Walsh codes as used in IS-95 are a set of 64 binary orthogonal sequences. These sequences are orthogonal to each other, and they are generated by using the Hadamard matrix. Recursion is used to generate higher order matrices from lower order ones; that is,

$$\mathbf{H}_{2N} = \begin{bmatrix} \mathbf{H}_N & \mathbf{H}_N \\ \mathbf{H}_N & \overline{\mathbf{H}_N} \end{bmatrix} \tag{3.6}$$

where $\overline{\mathbf{H}_N}$ contains the same but inverted elements of $\mathbf{H}_N$. The seed matrix is

$$\mathbf{H}_2 = \begin{bmatrix} 0 & 0 \\ 0 & 1 \end{bmatrix} \tag{3.7}$$

Therefore, to derive a set of four orthogonal Walsh sequences $\mathbf{w}_0$, $\mathbf{w}_1$, $\mathbf{w}_2$, and $\mathbf{w}_3$, we only need to generate a Hadamard matrix of order 4, or

$$\mathbf{H}_4 = \begin{bmatrix} \mathbf{H}_2 & \mathbf{H}_2 \\ \mathbf{H}_2 & \overline{\mathbf{H}_2} \end{bmatrix} = \begin{bmatrix} 0 & 0 & 0 & 0 \\ 0 & 1 & 0 & 1 \\ 0 & 0 & 1 & 1 \\ 0 & 1 & 1 & 0 \end{bmatrix}$$

The four orthogonal sequences in this Walsh code set are taken from the rows of the matrix $\mathbf{H}_4$; that is,

$$\mathbf{w}_0 = \begin{bmatrix} 0 & 0 & 0 & 0 \end{bmatrix}$$
$$\mathbf{w}_1 = \begin{bmatrix} 0 & 1 & 0 & 1 \end{bmatrix}$$
$$\mathbf{w}_2 = \begin{bmatrix} 0 & 0 & 1 & 1 \end{bmatrix}$$
$$\mathbf{w}_3 = \begin{bmatrix} 0 & 1 & 1 & 0 \end{bmatrix}$$

For DS-SS multiple access, Section 1.2 specifies three conditions that must be met by a set of orthogonal sequences. The three conditions are

1. The cross-correlation should be zero or very small.
2. Each sequence in the set has an equal number of 1s and −1s, or the number of 1s differs from the number of −1s by at most one.
3. The scaled dot product of each code should equal to 1.

By changing the 0s to −1s in each of the four sequences above, that is,

$$\mathbf{w}_0 = \begin{bmatrix} -1 & -1 & -1 & -1 \end{bmatrix}$$
$$\mathbf{w}_1 = \begin{bmatrix} -1 & +1 & -1 & +1 \end{bmatrix}$$
$$\mathbf{w}_2 = \begin{bmatrix} -1 & -1 & +1 & +1 \end{bmatrix}$$
$$\mathbf{w}_3 = \begin{bmatrix} -1 & +1 & +1 & -1 \end{bmatrix}$$

we can facilitate the calculation of cross products and dot products. The readers can easily verify that all of the above sequences except $\mathbf{w}_0$ satisfy the conditions. In general, the 0th Walsh sequence consists of all −1s and thus cannot be used for channelization. In the IS-95 CDMA system, $\mathbf{w}_0$ is not used to transmit any baseband information.

Equation (3.6) can be recursively used to generate Hadamard matrices of higher orders in order to obtain larger sets of orthogonal sequences. For example, 8 orthogonal sequences, each of length 8, can be obtained by generating $\mathbf{H}_8$; 16 orthogonal sequences, each of length 16, can be obtained by generating $\mathbf{H}_{16}$. The IS-95 forward link uses a set of 64 orthogonal Walsh sequences, thus the physical limitation on the number of channels on the forward link is 63 because in an IS-95 system, $\mathbf{w}_0$ is not used to transmit any baseband information.

Example 3.1

Equation (3.6) can be used to generate $\mathbf{H}_8$, which is

$$
\mathbf{H}_8 = \begin{bmatrix}
0 & 0 & 0 & 0 & 0 & 0 & 0 & 0 \\
0 & 1 & 0 & 1 & 0 & 1 & 0 & 1 \\
0 & 0 & 1 & 1 & 0 & 0 & 1 & 1 \\
0 & 1 & 1 & 0 & 0 & 1 & 1 & 0 \\
0 & 0 & 0 & 0 & 1 & 1 & 1 & 1 \\
0 & 1 & 0 & 1 & 1 & 0 & 1 & 0 \\
0 & 0 & 1 & 1 & 1 & 1 & 0 & 0 \\
0 & 1 & 1 & 0 & 1 & 0 & 0 & 1
\end{bmatrix}
$$

The eight resulting orthogonal Walsh codes are

$$
\begin{aligned}
\mathbf{w}_0 &= \begin{bmatrix} -1 & -1 & -1 & -1 & -1 & -1 & -1 & -1 \end{bmatrix} \\
\mathbf{w}_1 &= \begin{bmatrix} -1 & +1 & -1 & +1 & -1 & +1 & -1 & +1 \end{bmatrix} \\
\mathbf{w}_2 &= \begin{bmatrix} -1 & -1 & +1 & +1 & -1 & -1 & +1 & +1 \end{bmatrix} \\
\mathbf{w}_3 &= \begin{bmatrix} -1 & +1 & +1 & -1 & -1 & +1 & +1 & -1 \end{bmatrix} \\
\mathbf{w}_4 &= \begin{bmatrix} -1 & -1 & -1 & -1 & +1 & +1 & +1 & +1 \end{bmatrix} \\
\mathbf{w}_5 &= \begin{bmatrix} -1 & +1 & -1 & +1 & +1 & -1 & +1 & -1 \end{bmatrix} \\
\mathbf{w}_6 &= \begin{bmatrix} -1 & -1 & +1 & +1 & +1 & +1 & -1 & -1 \end{bmatrix} \\
\mathbf{w}_7 &= \begin{bmatrix} -1 & +1 & +1 & -1 & +1 & -1 & -1 & +1 \end{bmatrix}
\end{aligned}
$$

3.5.1.2　Channelization Using Walsh Codes

The following example illustrates how Walsh codes can be used for DS-SS multiple access. Suppose that there are three different users, and each user wishes to send a separate message. The separate messages are

$$
\mathbf{m}_1 = \begin{bmatrix} +1 & -1 & +1 \end{bmatrix} \qquad \mathbf{m}_2 = \begin{bmatrix} +1 & +1 & -1 \end{bmatrix} \qquad \mathbf{m}_3 \begin{bmatrix} -1 & +1 & +1 \end{bmatrix}
$$

Each of the three users is assigned a Walsh code, respectively:

$$\mathbf{w}_1 = \begin{bmatrix} -1 & +1 & -1 & +1 \end{bmatrix}$$

$$\mathbf{w}_2 = \begin{bmatrix} -1 & -1 & +1 & +1 \end{bmatrix}$$

$$\mathbf{w}_3 = \begin{bmatrix} -1 & +1 & +1 & -1 \end{bmatrix}$$

Each message is spread by its assigned Walsh code. Note that the chip rate of the Walsh code is four times the bit rate of the message, contributing to a processing gain of 4. For message one:

$m_1(t)$	1				−1				1			
$m_1(t)$	1	1	1	1	−1	−1	−1	−1	1	1	1	1
$w_1(t)$	−1	1	−1	1	−1	1	−1	1	−1	1	−1	1
$m_1(t)w_1(t)$	−1	1	−1	1	1	−1	1	−1	−1	1	−1	1

Note that $m_1(t)w_1(t)$ is the spread-spectrum signal of the first message. Similarly, for message two:

$m_2(t)$	1				1				−1			
$m_2(t)$	1	1	1	1	1	1	1	1	−1	−1	−1	−1
$w_2(t)$	−1	−1	1	1	−1	−1	1	1	−1	−1	1	1
$m_2(t)w_2(t)$	−1	−1	1	1	−1	−1	1	1	1	1	−1	−1

For message three:

$m_3(t)$	−1				1				1			
$m_3(t)$	−1	−1	−1	−1	1	1	1	1	1	1	1	1
$w_3(t)$	−1	1	1	−1	−1	1	1	−1	−1	1	1	−1
$m_3(t)w_3(t)$	1	−1	−1	1	−1	1	1	−1	−1	1	1	−1

The spread-spectrum signals for all three messages, $m_1(t)w_1(t)$, $m_2(t)w_2(t)$, and $m_3(t)w_3(t)$, are combined to form a composite signal $C(t)$; that is,

$$C(t) = m_1(t)w_1(t) + m_2(t)w_2(t) + m_3(t)w_3(t)$$

The resulting $C(t)$ is

$C(t)$	−1	−1	−1	3	−1	−1	3	−1	−1	3	−1	−1

$C(t)$ is the composite signal that is transmitted in the single RF band. If there are negligible errors during the transmission process, the receiver intercepts $C(t)$. In order to separate out the original messages $m_1(t)$, $m_2(t)$, and $m_3(t)$ from the composite signal $C(t)$, the receiver multiplies $C(t)$ by the assigned Walsh code for each message:

$C(t)w_1(t)$	1	−1	1	3	1	−1	−3	−1	1	3	1	−1
$C(t)w_2(t)$	1	1	−1	3	1	1	3	−1	1	−3	−1	−1
$C(t)w_3(t)$	1	−1	−1	−3	1	−1	3	1	1	3	−1	1

Then the receiver integrates, or adds up, all the values over each bit period. The functions $M_1(t)$, $M_2(t)$, and $M_3(t)$ are the results:

$C(t)w_1(t)$	1	−1	1	3	1	−1	−3	−1	1	3	1	−1
$M_1(t)$				4				−4				4

$C(t)w_2(t)$	1	1	−1	3	1	1	3	−1	1	−3	−1	−1
$M_2(t)$				4				4				−4

$C(t)w_3(t)$	1	−1	−1	−3	1	−1	3	1	1	3	−1	1
$M_3(t)$				−4				4				4

A "decision threshold" looks at the integrated functions $M_1(t)$, $M_2(t)$, and $M_3(t)$. The decision rules used are

$$\tilde{m}(t) = 1 \quad \text{if } M(t) > 0$$
$$\tilde{m}(t) = -1 \quad \text{if } M(t) < 0$$

After applying the above decision rules, we obtain the results:

$\tilde{m}_1(t)$	1	−1	1
$\tilde{m}_2(t)$	1	1	−1
$\tilde{m}_3(t)$	−1	1	1

3.5.1.3 Concluding Remarks

We have just illustrated how orthogonal Walsh codes can be used to provide channelization of different users. However, the ability to channelize depends heavily on the orthogonality of the code sequences during *all* stages of the transmission. For example, if due to multipath delay one of the users' codes is delayed by one chip, then the delayed code is no longer orthogonal to the other (nondelayed) codes in the code set. For example, the two Walsh codes

$$\mathbf{w}_2 = \begin{bmatrix} -1 & -1 & +1 & +1 \end{bmatrix}$$
$$\mathbf{w}_3 = \begin{bmatrix} -1 & +1 & +1 & -1 \end{bmatrix}$$

are orthogonal. However, if $\mathbf{w}_3$ is delayed by one chip, that is,

$$\mathbf{w'}_3 = \begin{bmatrix} -1 & -1 & +1 & +1 \end{bmatrix}$$

then the reader can easily verify that $\mathbf{w}_2$ and $\mathbf{w'}_3$ are no longer orthogonal. Therefore, synchronization is essential for using Walsh codes for DS-SS multiple access. In practice, the IS-95 CDMA system uses a pilot channel and a sync channel to synchronize the forward link and to ensure that the link is coherent.

3.5.2 PN Codes

Although the forward link of IS-95 CDMA has pilot and sync channels to aid synchronization, the reverse link does not have pilot and sync channels. The mobile stations transmit at will, and no attempt is made to synchronize their transmissions. Thus Walsh codes cannot be used on the reverse link. The incoherent nature of the reverse link calls for the use of another class of codes, PN codes, for channelization.

3.5.2.1 Generation of PN Codes

PN code sets can be generated from linear feedback shift registers. One such example (a three-stage register) is shown in Figure 3.10. Binary bits are shifted through the different stages of the register. The output of the last stage and the output of one intermediate stage are combined and fed as input to the first stage. The register starts with an initial sequence of bits, or initial state, stored in its stages. Then the register is clocked, and bits are moved through the stages. This way, the register continues to generate output bits and feed input bits to its first stage.

The output bits of the last stage form the PN code. We now demonstrate the code generation using the register shown in Figure 3.10. An initial state of [1, 0, 1] is used for the register. The output of stage 3 is the output of the register. After clocking the bits through the register, we obtain the results summarized in Table 3.2.

Note that at shift 7, the state of the register returns to that of the initial state, and further shifting of the bits yields another identical sequence of

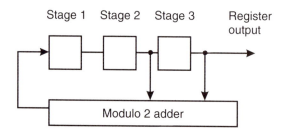

Figure 3.10 An example of linear feedback shift register for PN code generation.

outputs. Thus, the effective length of the periodic PN code generated is 7. The register output forms the PN code, which is

$$\mathbf{p} = \begin{bmatrix} 1 & 0 & 1 & 1 & 1 & 0 & 0 \end{bmatrix}$$

The code generated in this manner is called a maximal-length shift register code, and the length L of a maximal-length code is

$$L = 2^N - 1 \tag{3.8}$$

where N is the number of stages, or order, of the register. In this case, $N = 3$, and the code length equals 7. The PN code structure is determined by the feedback logic (i.e., which stages are tapped for feedback) and the initial register

Table 3.2
Register States and Outputs

Shift	Output Stage 1	Output Stage 2	Output Stage 3	Output Register
0	1	0	1	1
1	1	1	0	0
2	1	1	1	1
3	0	1	1	1
4	0	0	1	1
5	1	0	0	0
6	0	1	0	0
7	1	0	1	1

state. For example, if the initial state of the register is [0, 0, 0], then the different stages would get "stuck" in zeros; the register output then would be all zeros, and the code generated would not be maximal length.

A PN code set of seven codes can be generated by successively shifting **p**, and by changing 0s to −1s we obtain

$$\mathbf{p}_0 = \begin{bmatrix} +1 & -1 & +1 & +1 & +1 & -1 & -1 \end{bmatrix}$$
$$\mathbf{p}_1 = \begin{bmatrix} -1 & +1 & -1 & +1 & +1 & +1 & -1 \end{bmatrix}$$
$$\mathbf{p}_2 = \begin{bmatrix} -1 & -1 & +1 & -1 & +1 & +1 & +1 \end{bmatrix}$$
$$\mathbf{p}_3 = \begin{bmatrix} +1 & -1 & -1 & +1 & -1 & +1 & +1 \end{bmatrix}$$
$$\mathbf{p}_4 = \begin{bmatrix} +1 & +1 & -1 & -1 & +1 & -1 & +1 \end{bmatrix}$$
$$\mathbf{p}_5 = \begin{bmatrix} +1 & +1 & +1 & -1 & -1 & +1 & -1 \end{bmatrix}$$
$$\mathbf{p}_6 = \begin{bmatrix} -1 & +1 & +1 & +1 & -1 & -1 & +1 \end{bmatrix}$$

Readers can easily verify that these codes satisfy the conditions outlined in Section 1.2 for DS-SS multiple access; that is,

1. The cross-correlation should be zero or very small.
2. Each sequence in the set has an equal number of 1s and −1s, or the number of 1s differs from the number of −1s by at most one.
3. The scaled dot product of each code should equal to 1.

Since the maximal PN code length is always an odd number (see (3.8)) and the code shown above has four +1s and three −1s, the code satisfies condition 2.

3.5.2.2 Channelization using PN Codes

We again use an example to illustrate how PN codes can be used for multiple access. Suppose the same three users wish to send three separate messages. These messages are

$$\mathbf{m}_1 = \begin{bmatrix} +1 & -1 & +1 \end{bmatrix}$$
$$\mathbf{m}_2 = \begin{bmatrix} +1 & +1 & -1 \end{bmatrix}$$
$$\mathbf{m}_3 = \begin{bmatrix} -1 & +1 & +1 \end{bmatrix}$$

Each of the three users is assigned a PN code, respectively:

$$P_0 = \begin{bmatrix} +1 & -1 & +1 & +1 & +1 & -1 & -1 \end{bmatrix}$$
$$P_3 = \begin{bmatrix} +1 & -1 & -1 & +1 & -1 & +1 & +1 \end{bmatrix}$$
$$P_6 = \begin{bmatrix} -1 & +1 & +1 & +1 & -1 & -1 & +1 \end{bmatrix}$$

Message one is assigned PN code 0, message two is assigned PN code 3, and message three is assigned PN code 6. Each message is spread by its assigned PN code. Note that the chip rate of the PN code is seven times the bit rate of the message, contributing to a processing gain of 7. For message one:

```
m₁(t)        1                          -1                          1
m₁(t)      1  1  1  1  1  1  1 -1 -1 -1 -1 -1 -1 -1  1  1  1  1  1  1  1
p₀(t)      1 -1  1  1  1 -1 -1  1 -1  1  1  1 -1 -1  1 -1  1  1  1 -1 -1
m₁(t)p₀(t) 1 -1  1  1  1 -1 -1 -1  1 -1 -1 -1  1  1  1 -1  1  1  1 -1 -1
```

Note that $m_1(t)p_0(t)$ is the spread-spectrum signal of the first message. Similarly for message two:

```
m₂(t)        1                           1                         -1
m₂(t)      1  1  1  1  1  1  1  1  1  1  1  1  1  1 -1 -1 -1 -1 -1 -1 -1
p₃(t)      1 -1 -1  1 -1  1  1  1 -1 -1  1 -1  1  1  1 -1 -1  1 -1  1  1
m₂(t)p₃(t) 1 -1 -1  1 -1  1  1  1 -1 -1  1 -1  1  1 -1  1  1 -1  1 -1 -1
```

For message three:

```
m₃(t)       -1                           1                          1
m₃(t)     -1 -1 -1 -1 -1 -1 -1  1  1  1  1  1  1  1  1  1  1  1  1  1  1
p₆(t)     -1  1  1  1 -1 -1  1 -1  1  1  1 -1 -1  1 -1  1  1  1 -1 -1  1
m₃(t)p₆(t) 1 -1 -1 -1  1  1 -1 -1  1  1  1 -1 -1  1 -1  1  1  1 -1 -1  1
```

The spread-spectrum signals for all three messages $m_1(t)p_0(t)$, $m_2(t)p_3(t)$, and $m_3(t)p_6(t)$ are combined to form a composite signal $C(t)$; that is,

$$C(t) = m_1(t)p_0(t) + m_2(t)p_3(t) + m_3(t)p_6(t)$$

The resulting $C(t)$ is

$C(t)$ 3 –3 –1 1 1 1 –1 –1 1 –1 1 –3 1 3 –1 1 3 1 1 –3 –1

    $C(t)$ is the composite signal that is transmitted in the RF band. If there are negligible errors during the transmission process, the receiver intercepts $C(t)$. In order to separate out the original messages $m_1(t)$, $m_2(t)$, and $m_3(t)$ from the composite signal $C(t)$, the receiver multiplies $C(t)$ by the assigned PN code for each message:

$C(t)p_0(t)$ 3 3 –1 1 1 –1 1 –1 –1 –1 1 –3 –1 –3 –1 –1 3 1 1 3 1
$C(t)p_3(t)$ 3 3 1 1 –1 1 –1 –1 –1 1 1 3 1 3 –1 –1 –3 1 –1 –3 –1
$C(t)p_6(t)$ –3 –3 –1 1 –1 –1 –1 1 1 –1 1 3 –1 3 1 1 3 1 –1 3 –1

    Then the receiver integrates, or adds up, all the values over each bit period. The functions $M_1(t)$, $M_2(t)$, and $M_3(t)$ are the results:

$C(t)p_0(t)$ 3 3 –1 1 1 –1 1 –1 –1 –1 1 –3 –1 –3 –1 –1 3 1 1 3 1
$M_1(t)$ 7   –9   7

$C(t)p_3(t)$ 3 3 1 1 –1 1 –1 –1 –1 1 1 3 1 3 –1 –1 –3 1 –1 –3 –1
$M_2(t)$ 7   7   –9

$C(t)p_6(t)$ –3 –3 –1 1 –1 –1 –1 1 1 –1 1 3 –1 3 1 1 3 1 –1 3 –1
$M_3(t)$ –9   7   7

    A decision threshold looks at the integrated functions $M_1(t)$, $M_2(t)$, and $M_3(t)$. The decision rules used are

$$\tilde{m}(t) = 1 \quad \text{if } M(t) > 0$$
$$\tilde{m}(t) = -1 \quad \text{if } M(t) < 0$$

After applying the above decision rules, we obtain

$\tilde{m}_1(t)$ 1   –1   1
$\tilde{m}_2(t)$ 1   1   –1
$\tilde{m}_3(t)$ –1   1   1

3.5.2.3 Concluding Remarks

We define the discrete-time autocorrelation of a real-valued sequence **x** to be

$$R_x(i) = \sum_{j=0}^{J-1} x_j x_{j-1} \tag{3.9}$$

In other words, for each successive shift i, we calculate the summation of the product of x_j and its shifted version x_{j-i}. We proceed to calculate the autocorrelation of the PN sequence $\mathbf{p}_0$. Table 3.3 calculates the autocorrelation $R_{p_0}(i)$ of $\mathbf{p}_0$.

Note that $\mathbf{p}_0 = \begin{bmatrix} +1 & -1 & +1 & +1 & +1 & -1 & -1 \end{bmatrix}$ and the shifted sequence $\mathbf{p}_{0,j-i}$ are also shown for each shift i in the table. The resulting $R_{p_0}(i)$ for each shift i is shown on the right side of the table. Figure 3.11 depicts the autocorrelation function $R_{p_0}(i)$ as a function of time shift i.

In Figure 3.11, we see that the autocorrelation function reaches a peak at every seventh shift. For all other shifts (or time offsets), the autocorrelation stays at the minimum of −1. The autocorrelation property shown in

Table 3.3
Calculation of Autocorrelation for the Sequence **P₀**

i	$\mathbf{p}_{0,j-i}$							$R_{p_0}(i)$
0	1	−1	1	1	1	−1	−1	7
1	−1	1	−1	1	1	1	−1	−1
2	−1	−1	1	−1	1	1	1	−1
3	1	−1	−1	1	−1	1	1	−1
4	1	1	−1	−1	1	−1	1	−1
5	1	1	1	−1	−1	1	−1	−1
6	−1	1	1	1	−1	−1	1	−1
7	1	−1	1	1	1	−1	−1	7
8	−1	1	−1	1	1	1	−1	−1
9	−1	−1	1	−1	1	1	1	−1
10	1	−1	−1	1	−1	1	1	−1
11	1	1	−1	−1	1	−1	1	−1
12	1	1	1	−1	−1	1	−1	−1
13	−1	1	1	1	−1	−1	1	−1
14	1	−1	1	1	1	−1	−1	7
15	−1	1	−1	1	1	1	−1	−1

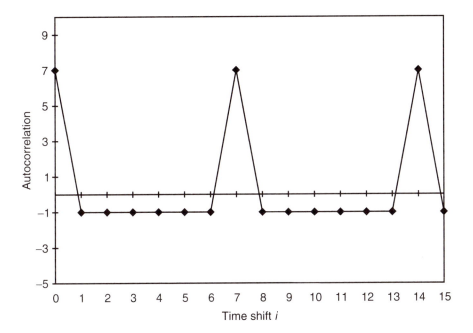

Figure 3.11 Autocorrelation function of PN sequence $\mathbf{p}_0$.

Figure 3.11 is important because it helps the initial acquisition and synchronization of the PN code at the receiver. A high correlation occurs only when the codes are aligned (i.e., when time shift i is zero); if the codes are not aligned, a low correlation results.

In practice, the receiver possesses an original copy of the PN code (i.e., $\mathbf{p}_{0,j}$). The receiver would like to acquire an incoming sequence $\mathbf{p}_{0,j-i}$ at an arbitrary phase. The receiver only has to "slide" the incoming sequence and calculate the autocorrelation. When the autocorrelation reaches a maximum, then the two codes are in-phase and have a time shift of zero. In an IS-95 CDMA system, this is in fact done by a mobile station to acquire the unmodulated pilot channel. This acquisition scheme can also be used when the spreading code length is equal to the data bit period.

In IS-95 CDMA, the reverse link uses the "long" PN code for channelization. The long code is so called because its length is literally very long. The long code has a length of $2^{42} - 1$ chips and is generated using a 42-stage register.

We saw in Section 3.5.1 that the forward link uses the Walsh code for channelizing individual users of a particular base station. However, the forward link also uses the PN code. Each base station is assigned a unique PN code that is superimposed on top of the Walsh code. This is done to provide isolation among the different base stations (or sectors); the isolation is necessary because each base station uses the same 64 Walsh code set. The PN code used on the forward link is called the "short" code. It is so called because its length is relatively short. The short code is generated using a 15-stage register and has a length of $2^{15} - 1$ chips.

3.6 Modulation

The digital bit stream has to be modulated onto an RF carrier in order for it to be transmitted. The modulated signal is then transmitted through space in the form of a propagating *electromagnetic* (EM) field.

One immediate question is why do we need to modulate the bit stream onto an RF carrier; why can't we just transmit the baseband signal through space to the desired designation. There are two answers to this question. First, the government regulatory agency (i.e., the FCC) specifies the frequency at which a particular service can transmit. Thus, not everyone can transmit at the baseband frequency. Second, in order to transmit at baseband, which is at a much lower frequency, the required antenna size would be enormous in order to allow an efficient coupling between the transmitter and free space. For example, in order to efficiently couple power to free space, the antenna size needs to be at least on the order of the wavelength. If one wishes to transmit a baseband signal at 9.6 kHz, the antenna size would be 31.25 km!

There is a difference between analog and digital modulation techniques. Readers may be familiar with analog modulation schemes such as *amplitude modulation* (AM) and *frequency modulation* (FM). In analog modulation, information is contained in the continuous-waveform shape of the signal. Digital modulation schemes, on the other hand, are used to transmit discrete units of information called symbols, and the information may be contained in the amplitude (e.g., on-off keying), the phase (e.g., phase-shift keying), or the amplitude and phase (e.g., quadrature-amplitude modulation) of the signal.

3.6.1 Binary Phase-Shift Keying (BPSK)

3.6.1.1 Modulator

Let's first examine a *basic digital modulation scheme* called BPSK and its performance in a Gaussian noise environment. The concept is simple. Whenever the transmitter wants to send a +1, it will transmit a positive cosinusoid; whenever the transmitter wants to send a −1, it will transmit a negative cosinusoid. The analytic expression for BPSK is

$$+1: s_{+1}(t) = \sqrt{\frac{2E}{T}} \cos(2\pi ft) \qquad 0 < t < T \tag{3.10}$$

$$-1: s_{-1}(t) = \sqrt{\frac{2E}{T}} \cos(2\pi ft + \pi) = -\sqrt{\frac{2E}{T}} \cos(2\pi ft) \qquad 0 < t < T \tag{3.11}$$

where E is the energy per symbol, and T is the time duration of each symbol. From these expressions, we can see that the information is indeed stored in the phase of the modulated signals $s_{+1}(t)$ and $s_{-1}(t)$. If the transmitted information is 1, the modulated signal $s_{+1}(t)$ has a phase of 0. If the transmitted information is −1, then the modulated signal $s_{-1}(t)$ has a phase of π, or 180 degrees. Figure 3.12 shows what the modulated signals look like in the time domain.

The BPSK modulator is quite simple to implement. The modulator itself is no more than a multiplier. Figure 3.13 shows the block diagram of a BPSK modulator. The input to the modulator consists of the data symbols. The data can be either a +1 or a −1. The data is multiplied by the carrier $\cos(2\pi ft)$ scaled by the coefficient $\sqrt{2E/T}$. The output of the multiplier is the corresponding modulated signal.

3.6.1.2 Demodulator

One implementation of the BPSK demodulator is the matched-filter approach. Figure 3.14 illustrates such an implementation. The received signal $r(t)$ has two components: the originally transmitted signal $s_i(t)$ where i could be either +1 or −1; and noise $n(t)$, which has been introduced by the channel. The received signal $r(t)$ is multiplied by the reference signals $s_{+1}(t)$. The multiplied result is then integrated over one bit interval T.

If the transmitter sent a +1 (i.e., $s_i(t) = s_{+1}(t)$), then the integrated result is

$$y = \frac{2E}{T} \int_0^T \cos^2(2\pi ft)dt + \sqrt{\frac{2E}{T}} \int_0^T \cos(2\pi ft)n(t)dt \tag{3.12}$$

where the first term is the *signal* term that is utilized by the decision threshold to make the decision and the second term is the *noise* contribution. In the absence of noise, we see that the first term reduces to

$$\frac{2E}{T} \int_0^T \cos^2(2\pi ft)dt = \frac{2E}{T}\left(\frac{1}{2}\right) = +\frac{E}{T} \qquad \text{(for } +1)$$

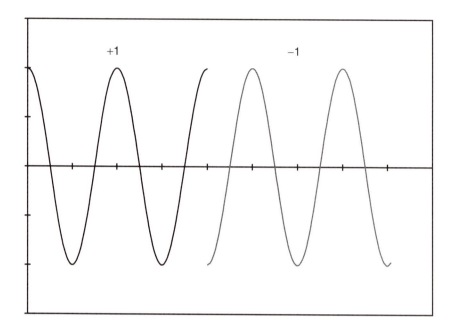

Figure 3.12 Modulated signal in the time domain using BPSK.

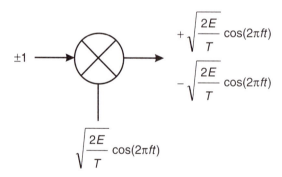

Figure 3.13 BPSK modulator.

If the transmitter sent a −1 (i.e., $s_i(t) = s_{-1}(t)$), then the integrated result is

$$y = -\frac{2E}{T}\int_0^T \cos^2(2\pi ft)\,dt - \sqrt{\frac{2E}{T}}\int_0^T \cos(2\pi ft)\,n(t)\,dt \qquad (3.13)$$

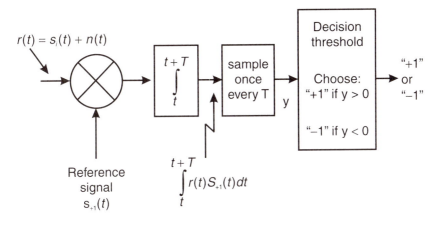

Figure 3.14 Correlator demodulator for BPSK.

In the absence of noise, we see that the first term reduces to

$$-\frac{2E}{T}\int_0^T \cos^2(2\pi ft)\,dt = -\frac{2E}{T}\left(\frac{1}{2}\right) = \frac{E}{T} \qquad (\text{for } -1)$$

Therefore, the decision threshold decides that the transmitter sent a +1 if the integrated result is greater than 0 and −1 if the integrated result is less than 0. In this *maximum likelihood detector* implementation, we have assumed that the probability of sending a +1 is equal to the probability of sending a −1. Furthermore, it is assumed that the demodulator is coherent (i.e., the phase of its reference signal perfectly matches the phase of the transmitter).

3.6.1.3 Error Performance of BPSK

Thus far, we have ignored the noise term at the output of the integrator. The noise terms (i.e., the second terms in (3.12) and (3.13)) are

$$\pm\sqrt{\frac{2E}{T}}\int_0^T \cos(2\pi ft)\,n(t)\,dt$$

The noise terms could be large enough to cause the decision threshold function to make a wrong decision. For example, suppose that the transmitter sent a +1. The output of the integrator is

$$y = \frac{2Ew}{T}\int_0^T \cos^2(2\pi ft)dt + \sqrt{\frac{2E}{T}}\int_0^T \cos(2\pi ft)n(t)dt$$

$$= +\left(\frac{E}{T}\right) + \sqrt{\frac{2E}{T}}\int_0^T \cos(2\pi ft)n(t)dt$$

The noise $n(t)$ is frequently modeled as an *additive white Gaussian noise* (AWGN) process. If the noise power is large (i.e., if the variance of $n(t)$ is large), it could cause the second term in the above expression to become less than $-(E/T)$. If that is the case, y would be less than zero and the decision threshold would then decide -1. Since the transmitter in reality sent a $+1$, the demodulator is said to have made an error.

To characterize the error performance of a coherent BPSK system, we use the signal-space representation of the BPSK signals in (3.10) and (3.11). The signal-space representation is nothing more than another representation of the signals. The representation depicts a signal in its in-phase and quadrature components. Every real-valued signal can be de-composed into an I and a Q component, and the signal-space representation effectively draws the signal in a space defined by the I and the Q axes. Figure 3.15 shows the signal-space representation of the BPSK signals defined in equations (3.10) and (3.11)

To obtain the magnitude of a signal along the I axis, we extract the in-phase component of that particular signal. The in-phase component is calculated by multiplying that signal by a cosinusoid and then integrating over the bit period. For example, the in-phase component of $s_{+1}(t)$ is

$$\int_0^T \cos(2\pi ft)s_{+1}(t)dt = \sqrt{\frac{2E}{T}}\int_0^T \cos^2(2\pi ft)dt = \sqrt{\frac{2E}{T}}\left(\frac{1}{2}\right) = \sqrt{\frac{E}{2T}} = \sqrt{\frac{E}{2}}$$

For simplicity, we use $T = 1$.

The probability that an error has occurred happens when the transmitter sent a $+1$ but the receiver made a decision of -1, and vice versa. In other words,

$$P_e = P\left(\text{decide} - 1 | \text{sent} + 1\right)P(\text{sent} + 1) + P\left(\text{decide} + 1 | \text{sent} - 1\right)P(\text{sent} - 1)$$

If we assume that the probability of sending a $+1$ is equal to the probability of sending a -1, then

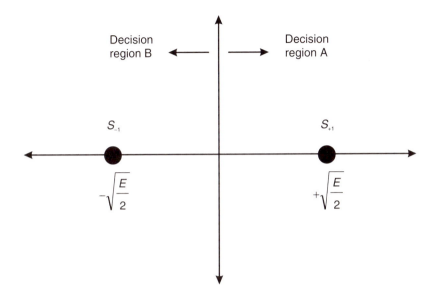

Figure 3.15 Signal-space representation of a BSPK signal set.

$$P_e = \frac{1}{2}P\left(\text{decide} - 1\big|\text{sent} + 1\right) + \frac{1}{2}P\left(\text{decide} + 1\big|\text{sent} - 1\right) \qquad (3.14)$$

The two probabilities in the above equation can be obtained by noting the fact that in its *IQ* form, the received signal *y* is given by $y = s + n$, where *s* is the signal and *n* is the noise. Note that *s* is a constant in the *IQ* representation, and *n* is effectively a Gaussian random variable; thus, *y* is also Gaussian distributed with a mean of *s*. Therefore, the probability of deciding −1 given that +1 is sent is the probability of *y* (given +1 is sent) falling in decision region B. This probability can be evaluated by integrating the Gaussian probability density function over the error area. See Figure 3.16.

Since the two conditional probability density functions are symmetrical, the two conditional probabilities are identical; that is,

$$P\left(\text{decide} - 1\big|\text{sent} + 1\right) = P\left(\text{decide} + 1\big|\text{sent} - 1\right)$$

and (3.14) reduces to

$$P_e = P\left(\text{decide} - 1\big|\text{sent} + 1\right) \qquad (3.15)$$

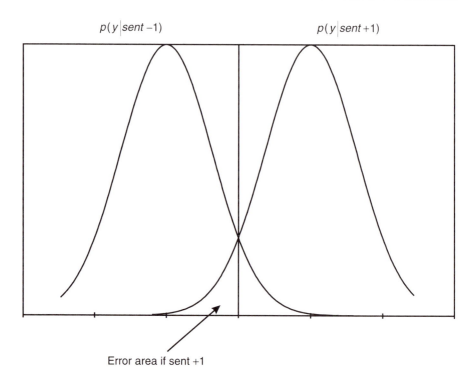

Figure 3.16 Conditional probability density function for the received signals.

Equation (3.15) can be evaluated as

$$P_e = \int_{-\infty}^{0} p\left(y \middle| \text{sent} + 1\right) dy \tag{3.16}$$

Since $p(y|\text{sent} + 1)$ is a Gaussian probability density function, we can evaluate (3.16) by using the complementary error function $Q(x)$:

$$Q(x) = \frac{1}{\sqrt{2\pi}} \int_{x}^{\infty} \exp\left(-\frac{v^2}{2}\right) dv \tag{3.17}$$

By substituting the appropriate variables, and by recognizing the fact that the noise variance σ^2 equates to half of the noise power density N_0 [1], we obtain

$$P_e = \frac{1}{2} Q\left(\sqrt{\frac{E}{N_0}} \right) \tag{3.18}$$

Since in BPSK each symbol is also an individual bit, the probability of symbol error shown in (3.18) is also equal to the probability of bit error. The energy per symbol E is also the energy per bit E_b:

$$P_b = \frac{1}{2} Q\left(\sqrt{\frac{E_b}{N_0}} \right) \tag{3.19}$$

Figure 3.17 shows the curve of probability of bit error versus E_b/N_0.

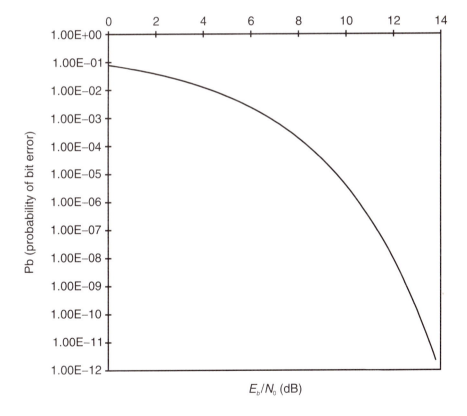

Figure 3.17 Bit error performance of a coherent BPSK system.

3.6.2 Quadrature Phase-Shift Keying (QPSK)

BPSK is capable of transmitting one bit of information (+1 or −1) per symbol period T. In this section, we examine another modulation scheme that can transmit two bits of information per symbol period. *Quadrature phase-shift keying* (QPSK) makes use of the quadrature component in addition to the in-phase component. The in-phase and quadrature components can be combined without interfering with each other because the two are orthogonal to each other; that is,

$$\int_0^{k/f} \cos(2\pi f t) \sin(2\pi f t) dt = 0 \qquad k = 0,1,2,\ldots$$

Therefore, a second BPSK signal in quadrature can be added to the first without introducing interference to either one. This technique, known as QPSK, effectively doubles the bandwidth efficiency of BPSK because it is able to transmit an additional bit during T.

3.6.2.1 Modulator

In order to send two bits of information, the QPSK system needs to use four symbols: symbols 0, 1, 2, and 3. For example, if the transmitter wants to transmit two bits {−1,−1} during T, it would transmit symbol 0. If the transmitter wants to transmit two bits {−1,−1} during T, it would transmit symbol 2, and so on. The mapping shown in Table 3.4 is used.

In order to transmit four different symbols, the QPSK transmitter needs to be able to send four different waveforms: $s_0(t)$, $s_1(t)$, $s_2(t)$, and $s_3(t)$. Each signal corresponds to one of the four symbols; that is,

Table 3.4
Mapping Between Transmitted Symbols and Represented Bits

Transmitted Symbol	Represented Bits
0	+1, +1
1	−1, +1
2	−1, −1
3	+1, −1

$$\text{Symbol 0: } s_0(t) = \sqrt{\frac{2E}{T}} \cos(2\pi ft + \pi/4) \qquad 0 < t < T \qquad (3.20)$$

$$\text{Symbol 1: } s_1(t) = \sqrt{\frac{2E}{T}} \cos(2\pi ft + (3\pi/4))$$

$$= \sqrt{\frac{2E}{T}} \sin(2\pi ft + \pi/4) \qquad 0 < t < T \qquad (3.21)$$

$$\text{Symbol 2: } s_2(t) = \sqrt{\frac{2E}{T}} \cos(2\pi ft + (5\pi/4))$$

$$= -\sqrt{\frac{2E}{T}} \cos(2\pi ft + \pi/4) \qquad 0 < t < T \qquad (3.22)$$

$$\text{Symbol 3: } s_3(t) = \sqrt{\frac{2E}{T}} \cos(2\pi ft + (7\pi/4)) \qquad (3.23)$$

$$= -\sqrt{\frac{2E}{T}} \sin(2\pi ft + \pi/4) \qquad 0 < t < T$$

where E is the energy per symbol. The transmitter changes the phase of the cosine waveform, depending on which symbol is to be transmitted. The phase can change to any one of the four states: 45, 135, 225, and 315 degrees. Figure 3.18 shows QPSK's signal constellation in signal space.

Figure 3.19 shows the block diagram of a QPSK modulator. The input to the modulator are the data bits $\{a_n\}$; $n = 0,1,2,3,\ldots$. The data bit can be either a +1 or a −1. The data bits are fed into a *demultiplexer* (DEMUX), where the bit stream is separated into an even bit stream and an odd bit stream. The even bit stream is multiplied by the in-phase carrier, and the odd bit stream is multiplied by the quadrature carrier. The output of the multipliers is combined in the adder to form the QPSK signal.

3.6.2.2 Demodulator

The QPSK receiver can use the same matched-filter and maximum likelihood detector approach, except in this case another branch needs to be added for the quadrature component. Figure 3.20 illustrates such an implementation. The received signal $r(t)$ is fed into two separate paths: the in-phase path and the

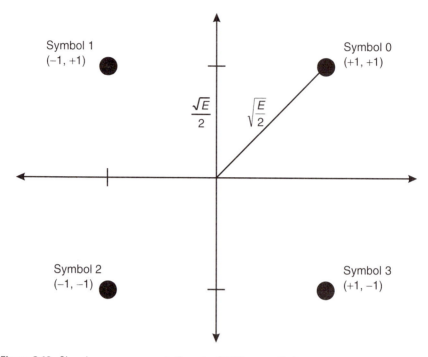

Figure 3.18 Signal-space representation of a QPSK constellation.

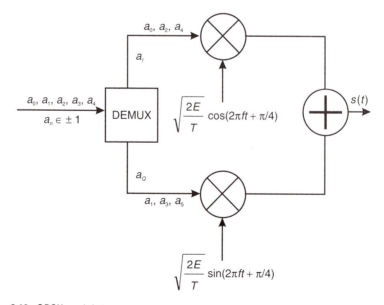

Figure 3.19 QPSK modulator.

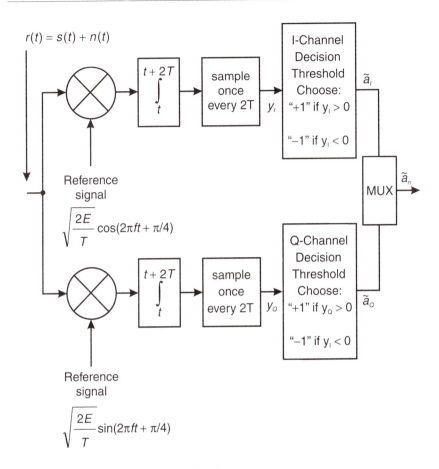

Figure 3.20 Correlator demodulator for QPSK.

quadrature path, where the signals are multiplied by the respective reference signals. The multiplied results are integrated and fed into the decision threshold. The decision rule is the same as that of the BPSK system, namely decide +1 if y is greater than zero and decide −1 if y is less than zero. The resulting decisions are multiplexed by the *multiplexer* (MUX) to form the recovered data bit stream $\tilde{a}_n$.

Note that in this correlator implementation, we have again assumed that the probability of sending a +1 is equal to the probability of sending a −1. Furthermore, it is assumed that the demodulator is coherent, which means that the phases of the reference signals perfectly match the phases of the modulating carriers.

3.6.2.3 Error Performance of QPSK

Using the P_b result obtained for BPSK, we can easily derive the probability of error for a QPSK system. Figure 3.21 shows the necessary conditions under which an error occurs. The noise variance (i.e., noise power) has to be such that it pushes symbol 0 into either decision region B or decision region D. An error would also occur if symbol 0 is pushed into decision region C; however, this case is considered less likely because the noise power needed is greater than that of the two previously mentioned cases.

Let's first examine the probability P that symbol 0 will be pushed into decision region B by noise. Recall that the probability of bit error for BPSK is

$$P_{b,\text{BPSK}} = \frac{1}{2} Q\left(\sqrt{\frac{E}{N_0}} \right) \tag{3.24}$$

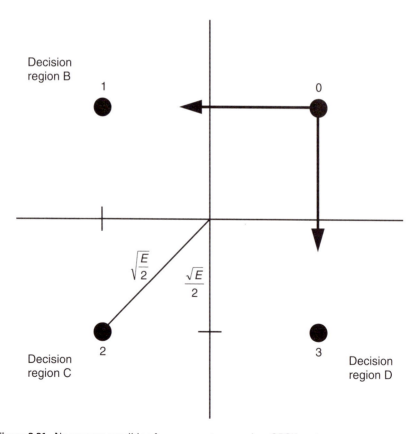

Figure 3.21 Necessary condition for an error to occur in a QPSK system.

Note that $P_{b,\text{BPSK}}$ is the probability that a BPSK symbol has been pushed a distance of $\sqrt{E/2}$ (see Figure 3.15). Therefore, we can express (3.24) in terms of an error distance d in signal space:

$$P_{b,\text{BPSK}} = \frac{1}{2} Q\left(\sqrt{\frac{2d}{N_0}} \right)$$

where d is the distance that the symbol has to traverse in order for an error to occur. For the case of QPSK, we are interested in the probability that a symbol is pushed by a distance of $\sqrt{E}/2$ (see Figure 3.21). Substituting $\sqrt{E}/2$ into d yields

$$P = \frac{1}{2} Q\left(\frac{1}{2}\sqrt{\frac{2E}{N_0}} \right) = \frac{1}{2} Q\left(\sqrt{\frac{E}{2N_0}} \right)$$

But E in the above expression is the energy per symbol. Since in QPSK each symbol contains two bits, energy per bit E_b is half of energy per symbol E_s; that is,

$$E_b = \frac{E_s}{2} \text{ for QPSK}$$ (3.25)

Substituting (3.25) into the previous expression for P, we get

$$P = \frac{1}{2} Q\left(\sqrt{\frac{E}{2N_0}} \right) = \frac{1}{2} Q\left(\sqrt{\frac{E_s}{2N_0}} \right) = \frac{1}{2} Q\left(\sqrt{\frac{E_b}{N_0}} \right)$$

P is the probability that symbol 0 would drift into decision region B. By symmetry, the probability that symbol 0 would drift into decision D is also P. The probability that symbol 0 would be in error is thus equal to the sum of the two probabilities since the occurrence of either of the two events would trigger an error. Therefore, the probability of symbol error P_e is

$$P_e = P + P = 2P = 2\left(\frac{1}{2} Q\left(\sqrt{\frac{E_b}{N_0}} \right) \right) = Q\left(\sqrt{\frac{E_b}{N_0}} \right)$$ (3.26a)

A more exact expression for P_e is

$$P_e = Q\left(\sqrt{\frac{E_b}{N_0}}\right) - \frac{1}{4}Q^2\left(\sqrt{\frac{E_b}{N_0}}\right) \tag{3.26b}$$

We are also interested in the probability of bit error P_b, Note from Table 3.4 that when a symbol transitions to its nearest neighbor, there is at most one out of the two bits that changed sign. To state this differently, only one of the two bits changes when a symbol transitions to its nearest neighbor. Therefore, whenever a symbol error occurs, only one of the two bits contained in that symbol is also in error. This means that for QPSK, the probability of bit error is half of the probability of symbol error; that is,

$$P_b = \frac{1}{2}P_e = \frac{1}{2}Q\left(\sqrt{\frac{E_b}{N_0}}\right) \tag{3.27}$$

From (3.27), we see that BPSK and QPSK have the same probability of bit error as a function of E_b/N_0.

3.6.3 Applications in IS-95 CDMA System

The IS-95 CDMA system uses QPSK for both forward and reverse links. The reverse link, in particular, uses a variant of QPSK called *offset quadrature phase-shift keying* (OQPSK). OQPSK differs from the conventional QPSK in that prior to carrier multiplication, a delay of a half-bit interval (with respect to the *I* path) is placed in the *Q* path (see Figure 3.22). This is done to avoid a 180-degree phase transition that occurs in conventional QPSK systems. For example, when symbol 0 transitions to symbol 3, the signal goes through a 180-degree phase transition through the origin. In the time domain, the signal envelope collapses and momentarily reaches zero. This *zero crossing* demands a lot of dynamic range from the power amplifier. Thus, OQPSK is used on the reverse link where the power amplifier of the mobile is limited both in size and in performance. The extra delay of half a bit in the *Q* path ensures that there will be no direct transition between symbols 0 and 2 and between symbols 1 and 3, and thus no zero crossing.

It is also important to note that the error performance (i.e., probability of bit error) presented in this chapter is derived using an uncoded AWGN channel. In reality, error-correcting codes are used to improve the error

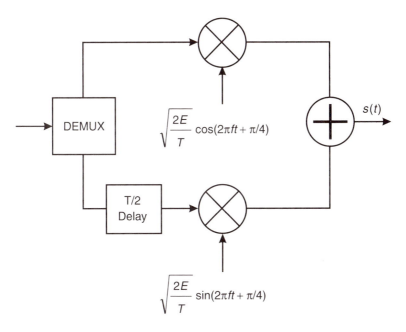

Figure 3.22 OQPSK modulator.

performance. For a fixed probability of error, error coding decreases the E_b/N_0 required to achieve that probability of error. In addition, the mobile communications environment is characterized by fading channels rather than the AWGN channel. The land-mobile fading channel typically increases the E_b/N_0 required to achieve a desired probability of error. System designers often need to resort to link-level simulations to assess the trade-off between E_b/N_0 and probability of error.

References

[1] Sklar, B., "A Structured Overview of Digital Communications," *IEEE Communication Magazine*, Aug. 1993.

[2] Schroeder, M. R., and B. S. Atal, "Code-Excited Linear Prediction (CELP): High-Quality Speech at Very Low Bit Rates," *Proc. IEEE Conf. Acoustic, Speech, and Signal Processing*, 1985.

[3] Blahut, R. E., *Theory and Practice of Error Control Codes*, Reading, MA: Addison-Wesley, 1984.

[4] Sklar, B., *Digital Communications: Fundamentals and Applications*, Englewood Cliffs, NJ: Prentice Hall, 1988.

[5] Stallings, W., and R. Van Slyke, *Business Data Communications*, Upper Saddle River, NJ: Prentice Hall, 1998.

[6] Stallings, W., *Data and Computer Communications*, Upper Saddle River, NJ: Prentice Hall, 1997.

[7] Lin, S., and D. J. Costello, Jr., *Error Control Coding: Fundamentals and Applications*, Englewood Cliffs, NJ: Prentice Hall, 1983.

[8] TIA/EIA IS-95A, "Mobile Station-Base Station Compatibility Standard for Dual-Mode Wideband Spread Spectrum Cellular System," Telecommunications Industry Association.

Select Bibliography

Gagliardi, R. M., *Satellite Communications*, New York, NY: Van Nostrand Reinhold, 1991.

Haykin, S., *Communication Systems*, New York, NY: John Wiley & Sons, 1983.

Jayant, N. S., and P. Noll, *Digital Coding of Waveforms,* Englewood Cliffs, NJ: Prentice Hall, 1984.

Markel, J. D., and A. H. Gray, Jr., *Linear Prediction of Speech*, New York, NY: Springer-Verlag, 1976.

Papoulis, A., *Probability, Random Variables and Stochastic Processes*, New York, NY: McGraw-Hill, 1965.

Peterson, R. L., R. E. Ziemer, and D. E. Borth, *Introduction to Spread Spectrum Communications*, Englewood Cliffs, NJ: Prentice Hall, 1995.

Proakis, J. G., *Digital Communications*, New York, NY: McGraw-Hill, 1989.

Schafer, R. W., and L. R. Rabiner, "Design of Digital Filter Banks for Speech Analysis," *Bell Syst. Tech. J.,* Vol. 50, No. 10, Dec. 1971, pp. 3097–3115.

Simon, M. K., S. M. Hinedi, and W. C. Lindsey, *Digital Communication Techniques: Signal Design and Detection*, Englewood Cliffs, NJ: Prentice Hall, 1995.

TIA IS-665, "W-CDMA Air Interface Compatibility Standard for 1.85-1.99 GHz PCS Applications," Telecommunications Industry Association.

TIA/EIA IS-96A, "Speech Service Option Standard for Wideband Spread Spectrum Digital Cellular System," Telecommunications Industry Association.

Viterbi, A. J., and J. K. Omura, *Principles of Digital Communication and Coding*, New York, NY: McGraw-Hill, 1979.

Wozencraft, J. M., and I. M. Jacobs, *Principles of Communication Engineering*, Waveland Press, 1990.

4

Principles of Code Division Multiple Access

4.1 Introduction

CDMA is a scheme by which multiple users are assigned radio resources using DS-SS techniques. Although all users are transmitting in the same RF band, individual users are separated from each other via the use of orthogonal codes. The North American CDMA standard, or IS-95, specifies that each user conveys baseband information at 9.6 Kbps (Rate Set 1), which is the rate of the vocoder output. The rate of the final spread signal is 1.2288 Mcps, resulting in an RF bandwidth of approximately 1.25 MHz.

There can be many 1.25-MHz signals present in the same RF band. To a large degree, the performance of a CDMA system is interference-limited. This means that the *capacity* and *quality* of the system are limited by the amount of interference power present in the band. Capacity is defined as the total number of simultaneous users the system can support, and quality is defined as the perceived condition of a radio link assigned to a particular user; this perceived link quality is directly related to the probability of bit error, or *bit error rate* (BER). This chapter presents those characteristics of a CDMA system that need to be optimized in order to reduce interference and increase quality.

4.2 Capacity

While there are many models of CDMA capacity in the current literature, we present a description of CDMA system capacity using the amount of user interference in the band. The actual capacity of a CDMA cell depends on many different factors, such as receiver demodulation, power-control accuracy, and actual interference power introduced by other users in the same cell and in neighboring cells.

In digital communication, we are primarily interested in a link metric called E_b/N_0, or energy per bit per noise power density. Chapter 3 reviews the performance of different digital modulation schemes in terms of probability of bit error as a function of E_b/N_0. This quantity can be related to the conventional *signal-to-noise ratio* (SNR) by recognizing that energy per bit equates to the average modulating signal power allocated to each bit duration; that is,

$$E_b = ST \qquad (4.1)$$

where S is the average modulating signal power and T is the time duration of each bit. Notice that (4.1) is consistent with dimensional analysis, which states that energy is equivalent to power multiplied by time. We can further manipulate (4.1) by substituting the bit rate R, which is the inverse of bit duration T:

$$E_b = \frac{S}{R}$$

E_b/N_0 is thus

$$\frac{E_b}{N_0} = \frac{S}{RN_0} \qquad (4.2)$$

We further substitute the noise power density N_0, which is the total noise power N divided by the bandwidth W; that is,

$$N_0 = \frac{N}{W} \qquad (4.3)$$

Substituting (4.3) into (4.2) yields

$$\frac{E_b}{N_0} = \frac{S}{N}\frac{W}{R} \qquad (4.4)$$

Equation (4.4) relates the energy per bit E_b/N_0 to two factors: the signal-to-noise ratio S/N of the link and the ratio of transmitted bandwidth W to bit rate R. The ratio W/R is also known as the processing gain of the system.

Here, we consider the reverse-link capacity since in CDMA this is often the limiting link in terms of capacity. Reverse link is the mobile to base station link. We assume that the system possesses perfect power control, which means that the transmitted powers of all mobile users are actively controlled such that at the base station receiver, the received powers from all mobile users are equal. Based on this assumption, the SNR of one user can be written as

$$\frac{S}{N} = \frac{1}{M-1} \tag{4.5}$$

where M is the total number of users present in the band. This is so because the total interference power in the band is equal to the sum of powers from individual users. Figure 4.1 illustrates the principle behind (4.5). Note that (4.5) also ignores other sources of interference such as thermal noise.

We proceed to substitute (4.5) into (4.4), and the result is

$$\frac{E_b}{N_0} = \frac{1}{(M-1)} \frac{W}{R} \tag{4.6}$$

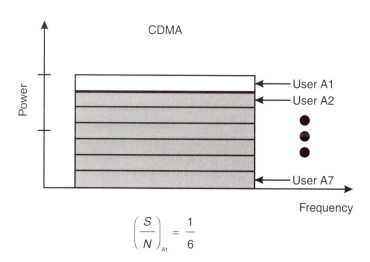

$$\left(\frac{S}{N}\right)_{A1} = \frac{1}{6}$$

Figure 4.1 In CDMA, the total interference power in the band is equal to the sum of powers from individual users. Therefore, if there are seven users occupying the band, and each user is power-controlled to the same power level, then the SNR experienced by any one user is 1/6.

Solving for $(M - 1)$ yields

$$M - 1 = \frac{(W/R)}{(E_b/N_0)} \tag{4.7}$$

Note that if M is large, then

$$M \approx \frac{(W/R)}{(E_b/N_0)} \tag{4.8}$$

4.2.1 Effects of Loading

Equation (4.8) is effectively a model that describes the number of users a single CDMA cell can support. This single cell is omnidirectional and has no neighboring cells, and the users are transmitting 100% of the time. In reality, there are many cells in a CDMA cellular or PCS system. Figure 4.2 shows that a particular cell (cell A) is bordered by other CDMA cells that are supporting other users. Although these other users from other cells are power-controlled by their respective home cells, the signal powers from these other users constitute interference to cell A. Therefore, cell A is said to be *loaded* by users from other cells. Equation (4.6) is modified to account for the effect of loading:

$$\frac{E_b}{N_0} = \frac{1}{(M - 1)} \frac{W}{R} \left(\frac{1}{1 + \eta}\right) \tag{4.9}$$

where η is the loading factor. η is a factor between 0% and 100%. In the example shown in Figure 4.3, the loading factor is 0.5 resulting in $(1 + 0.5)$, or a 150% increase of interference above those introduced by home users alone.

The inverse of the factor $(1 + \eta)$ is sometimes known as the *frequency reuse factor F*; that is,

$$F = \frac{1}{(1 + \eta)} \tag{4.10}$$

Note that the frequency reuse factor is ideally 1 in the single-cell case ($\eta = 0$). In the multicell case, as the loading η increases, the frequency reuse factor correspondingly decreases.

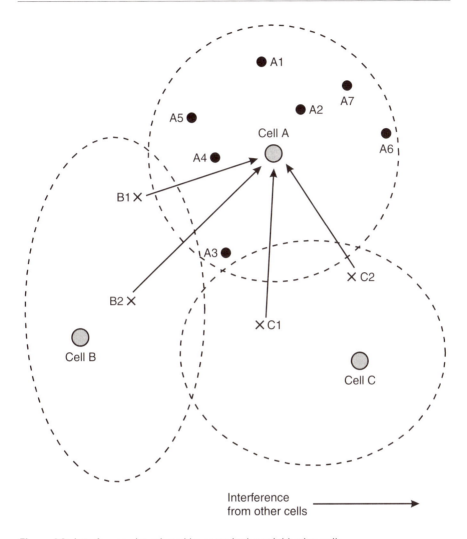

Figure 4.2 Interference introduced by users in the neighboring cell.

4.2.2 Effects of Sectorization

The interference from other users in other cells can be decreased if the cell in question is sectorized. Instead of having an omnidirectional antenna, which has an antenna pattern over 360 degrees, cell A can be sectorized to three sectors so that each sector is only receiving signals over 120 degrees (see Figure 4.4). In effect, a sectorized antenna rejects interference from users that are not within its antenna pattern. This arrangement decreases the effect of loading by a factor of

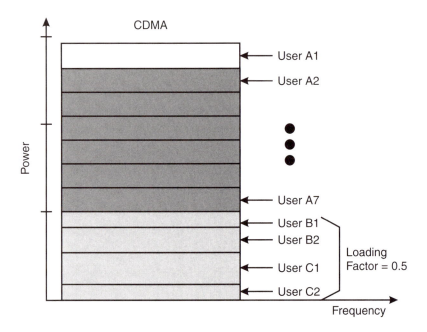

Figure 4.3 Loading factor as perceived by cell A.

approximately 3. If the cell is sectorized to six sectors, then the loading effect is decreased by a factor of approximately 6. This factor is called *sectorization gain* λ.

For one cell, the exact λ is obtained by dividing the total interference power from all directions by the perceived interference powers by the sector antenna; that is,

$$\lambda = \frac{\int_0^{2\pi} I(\theta)d\,\theta}{\int_0^{2\pi} \left(\frac{G(\theta)}{G(0)}\right) I(\theta)d\,\theta} \qquad (4.11)$$

where $G(\theta)$ is the horizontal antenna pattern of the sector antenna; $G(0)$ is the peak antenna gain, which is assumed to occur at boresight ($\theta = 0$); and $I(\theta)$ is the received interference power from users of other cells as a function of θ. The integrals in (4.11) are evaluated from 0 to 360 degrees. Equation (4.11) computes the exact sectorization gain, which depends heavily on the antenna gain of the antenna used, as well as on the spatial distribution and distance of

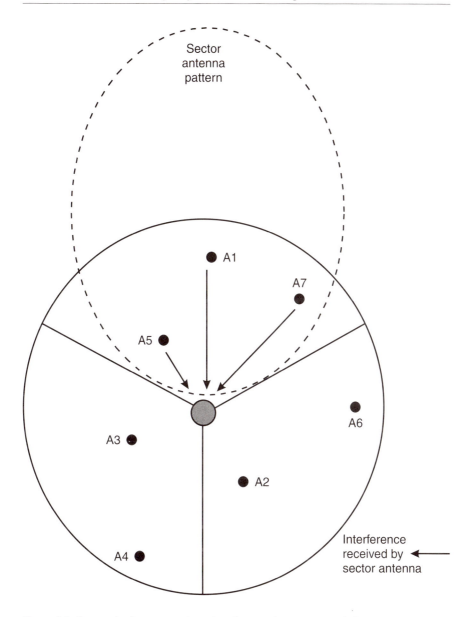

Figure 4.4 A sectorized antenna rejects interference from users not in its antenna pattern.

interfering users in other cells. Note that (4.11) does not take into account the vertical pattern of the sector antenna, the effect of which is quite small in calculating sectorization gain. In reality, λ is typically around 2.5 for three-sector configured systems and 5 for six-sector configured systems.

Equation (4.9) is thus modified to account for the effect of sectorization:

$$\frac{E_b}{N_0} = \frac{1}{(M-1)} \frac{W}{R} \left(\frac{1}{1+\eta}\right)\lambda \tag{4.12a}$$

4.2.3 Effects of Voice Activity

Equation (4.12a) assumes that all users are transmitting 100% of the time. In practice, the vocoder used by the IS-95 system is *variable rate*, which means that the output rate of the vocoder is adjusted according to a user's actual speech pattern. For example, if the user is not speaking during part of the conversation, the output rate of the vocoder is lowered to prevent power from being transmitted unnecessarily. The effect of this variable-rate vocoding is the reduction of overall transmitted power and hence interference. Speech statistics shows that a user in a conversation typically speaks between 40% and 50% of the time. By employing variable-rate vocoding, the system reduces the total interference power by this *voice activity factor*.

Thus, (4.12a) is again modified to account for the effect of voice activity:

$$\frac{E_b}{N_0} = \frac{1}{(M-1)} \frac{W}{R} \left(\frac{1}{1+\eta}\right)\lambda\left(\frac{1}{v}\right) \tag{4.12b}$$

where v is the voice activity factor. Note that the effect of voice activity is to reduce the denominator, or the interference portion of the equation.

Solving (4.12b) for M yields

$$M = 1 + \frac{(W/R)}{(E_b/N_0)}\left(\frac{1}{1+\eta}\right)\lambda\left(\frac{1}{v}\right) \tag{4.13}$$

If M is large, then

$$M \approx \frac{(W/R)}{(E_b/N_0)}\left(\frac{1}{1+\eta}\right)\lambda\left(\frac{1}{v}\right) \tag{4.14}$$

Examining (4.12b), we can draw several conclusions regarding CDMA capacity:

1. Capacity, or number of simultaneous users M, is directly proportional to the processing gain of the system.

2. The link requires a particular E_b/N_0 to attain an acceptable BER and ultimately an acceptable *frame error rate* (FER). Capacity is inversely proportional to the required E_b/N_0 of the link. The lower the required threshold E_b/N_0, the higher the system capacity.

3. Capacity can be increased if one can decrease the amount of loading from users in adjacent cells.

4. Spatial filtering, such as sectorization, increases system capacity. For example, a six-sector cell would have more capacity than a three-sector cell.

4.3 Power Control

4.3.1 Why Power Control?

Power control is essential to the smooth operation of a CDMA system. Because all users share the same RF band through the use of PN codes, each user looks like random noise to other users. The power of each individual user, therefore, must be carefully controlled so that no one user is unnecessarily interfering with others who are sharing the same band.

To illustrate how power control is essential in CDMA, we consider a single cell that has two hypothetical users (see Figure 4.5). We again examine the reverse-link case since this link is often the limiting link in CDMA. User 2 is much closer to the base station than user 1. If there is no power control, both

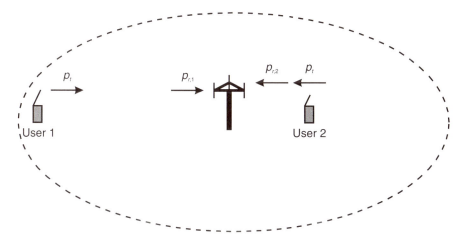

Figure 4.5 A base station with two hypothetical users. Each user is transmitting to the base station a fixed amount of power p_t.

users would transmit a fixed amount of power p_t; however, because of the difference in distance, the received power from user 2, or $p_{r,2}$, would be much larger than the received power from user 1, or $p_{r,1}$. If we assume that the difference in distance is such that $p_{r,2}$ is 10 times more than $p_{r,1}$, then user 1 would be at a great disadvantage.

If the required SNR, $(S/N)_{\text{required}}$, is (1/10), then we can immediately see the disparity between the SNRs of the two users. Figure 4.6 illustrates the point; if we ignore thermal noise, then the SNR of user 2, $(S/N)_2$, would be 10, and the SNR of user 1, $(S/N)_1$, would be (1/10). User 2 has a much higher SNR and thus enjoys great voice quality, but user 1's SNR is barely making the required $(S/N)_{\text{required}}$. This inequity is known as the classic *near-far* problem in a spread-spectrum multiple access system.

The system at this point is said to have reached its capacity. The reason is that if we attempt to add a third user transmitting p_t anywhere in the cell, then the SNR of that third user would not be able to reach the required $(S/N)_{\text{required}}$. Furthermore, if we force a third user onto the system, that third user not only will not attain the required $(S/N)_{\text{required}}$, but also will cause the SNR of user 2 to drop below the required $(S/N)_{\text{required}}$.

Power control is implemented to overcome the near-far problem and to maximize capacity. Power control is where the transmit power from each user is controlled such that the received power of each user at the base station is equal

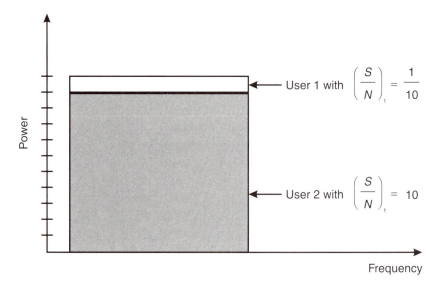

Figure 4.6 Received power from the two users at the base station. User 2 has a much higher SNR than user 1 does.

to one other. Figure 4.7 illustrates the concept. In the cell, if the transmit power of each user is controlled such that the received power of each user at the base station is equal to p_r, then a lot more users can be accommodated by the system. As a continuation of our previous example, if the required SNR $(S/N)_{\text{required}}$ is still (1/10), then a total of 11 users can be supported by the cell. The capacity is maximized with the use of power control (see Figure 4.8).

4.3.2 Reverse Link

4.3.2.1 Access Probes

One problem that has to be immediately solved in power control is the initial mobile transmit power. Before the mobile establishes contact with the base station, the mobile cannot be power-controlled by the base station. Thus, the natural question is when the mobile first attempts to access the base station, what power level should the mobile use to transmit its request? At this point, the base station has not yet made contact with the mobile user, and the base station has no idea as to the location of the mobile user. There are two options: the first option is that the mobile can attempt to access the base station with a high transmit power. Such high power increases the probability that the base station will receive that mobile's access request. However, the disadvantage of a high initial transmit power is that such high power represents interference to

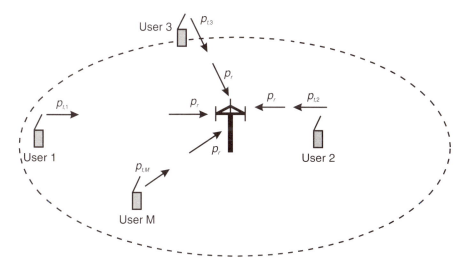

Figure 4.7 With power control, a base station can support many more users. Each user is power-controlled to transmit at different power levels. This is done so that the received powers of individual users are all equal at the base station.

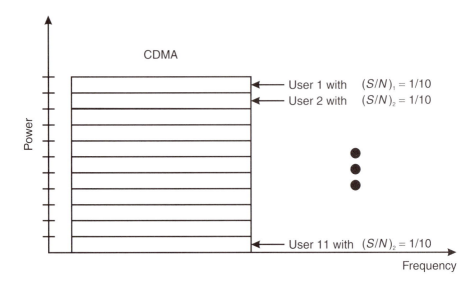

Figure 4.8 Capacity is maximized when the received powers of all users are equal at the base station.

other users currently served by the cell. The second option is that the mobile can request access from the base station with a low transmit power. Such low power decreases the likelihood that the base station will receive the mobile's access request. But the advantage is that this mobile won't cause much interference to other users.

The solution as specified in the IS-95 standard is that when the mobile first attempts to access the system, it transmits a series of *access probes*. Access probes are a series of transmissions of progressively higher power. The mobile transmits its first access probe at a relatively low power, then it waits for a response back from the base station. If after a random time interval the mobile does not receive an acknowledgment from the base station, then the mobile transmits a second access probe at a slightly higher power. The process repeats until the mobile receives an acknowledgment back from the base station. The power difference between the current access probe and the previous access probe is called an *access probe correction* (see Figure 4.9). The step size for a single access probe correction is specified by the system parameter PWR_STEP.

The standard further specifies that the mobile should use the power level it receives from the base station to estimate how much to initially transmit. In other words, if the mobile sees a strong signal from the base station, then it assumes that the base station is nearby and thus transmits initially at a relatively low level. If the mobile sees a weak signal from the base station, then it assumes

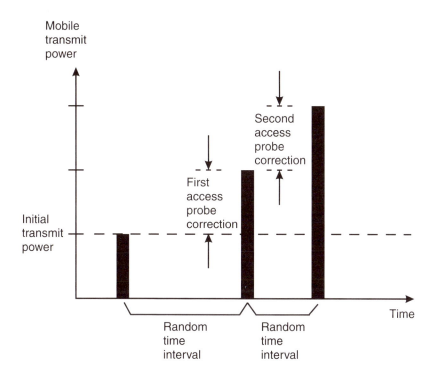

Figure 4.9 A series of access probes by the mobile to access the system. After [1].

that the base station is far away and thus transmits initially at a relatively high level. Knowing the received power from the base station, the mobile can estimate the forward path loss between the base station and itself. If it knows the transmit ERP of the base station, then the mobile would know how much it needs to transmit to compensate for that path loss. In reality, the mobile does not know the actual ERP of the base station, nor does it know how much received power is contributed by other, neighboring base stations. Therefore, a default constant is specified by the standard using generic assumptions of typical loading and base station ERPs. Specifically, the initial transmit power of the mobile, $p_{t,\text{initial}}$ in decibels, should be [1]

$$p_{t,\text{initial}} = -p_r - 73 + \text{NOM\_PWR} + \text{INIT\_PWR} \qquad (4.15)$$

As we can see, the default constant is −73 for cellular. A value of −76 is used for PCS systems. The two additional adjustments, NOM_PWR and INIT_PWR, can be set by the system operators for further fine-tuning. The values of these two adjustment factors, NOM_PWR and INIT_PWR, as well as the parameter PWR_STEP are broadcast by the base station (in the *access parameters message*) and received by the mobile prior to access probe transmission [1]. Upon receiving these two adjustment factors, the mobile uses them in (4.15) to determine its initial transmit power.

4.3.2.2 Open Loop

The process described above is termed *open-loop* power control in that it is purely a mobile-controlled operation and does not involve the base station at all. This open-loop process continues well after the base station has acknowledged the mobile's access request and after the mobile starts to transmit on a traffic channel.

After a call is established, and as the mobile moves around within the cell, the path loss between the mobile and the base station will continue to change. As a result, the received power at the mobile will change and the open-loop power control will continue to monitor the mobile received power p_r and adjust the mobile transmit power according to the following equation [1]:

$$p_t = -p_r - 73 + \text{NOM\_PWR} + \text{INIT\_PWR}$$
$$+ \text{ (sum of all access probe corrections)} \tag{4.16}$$

where p_t is the continuous open-loop estimate of the mobile transmit power. The difference between (4.16) and (4.15) is that (4.16) contains an additional term specifying the sum of all access probe corrections made during the access probe transmission.

It is important to note that the open-loop power control as specified in (4.16) is based on an estimate of the forward path loss. This power control is used to compensate for slow-varying and log-normal shadowing effects where there is a correlation between the forward-link and reverse-link fades. However, since the forward and reverse links are on different frequencies, the open-loop power control is inadequate and too slow to compensate for fast Rayleigh fading. Note that fast Rayleigh fading is frequency dependent and occurs over every half-wavelength (see Chapter 2 for a discussion of propagation phenomena in a mobile environment). In other words, since fast Rayleigh fading is frequency dependent, we cannot use open-loop power control (which assumes forward path loss is identical to reverse path loss) to compensate for fast Rayleigh fading.

4.3.2.3 Closed Loop

The *closed-loop* power control is used to compensate for power fluctuations due to fast Rayleigh fading. It is closed loop in that the process involves both the base station and the mobile. Once the mobile gets on a traffic channel and starts to communicate with the base station, the closed-loop power-control process operates along with the open-loop power control. In the closed-loop power control, the base station continuously monitors the reverse link and measures the link quality. If the link quality is getting bad, then the base station will command the mobile, via the forward link, to power up. If the link quality is too good, then there is excess power on the reverse link; in this case, the base station will command the mobile to power down. Ideally, FER is a good indicator of link quality. But because it takes a long time for the base station to accumulate enough bits to calculate FER, E_b/N_0 is used as an indicator of reverse link quality.

The reverse-link closed-loop power control is as follows:

1. The base station continuously monitors E_b/N_0 on the reverse link.

2. If E_b/N_0 is too high (i.e., if it exceeds a certain threshold [2]), then the base station commands the mobile to decrease its transmit power.

3. If E_b/N_0 is too low (i.e., if it drops below a certain threshold [2]), then the base station commands the mobile to increase its transmit power.

The base station sends the power-control commands to the mobile using the forward link. These power-control commands are in the form of *power-control bits* (PCBs). The amount of mobile power increase and power decrease per each PCB is nominally +1 dB and −1 dB.

Because the closed-loop power control is meant to combat fast Rayleigh fading, the mobile's response to these power-control commands must be very fast. For this reason, these PCBs are directly sent over the traffic channel. What actually happens is that bits are *robbed* from the traffic channel in order to send these PCBs. Figure 4.10 shows a simplified block diagram of a portion of the forward traffic channel generation.

The output from the vocoder and input into the convolutional encoder is 9.6 Kbps (at full rate for Rate Set 1). The Rate 1/2 convolutional encoder doubles the baseband rate to 19.2 Kbps. Prior to spreading, the PCBs at 800 bps are multiplexed onto the baseband stream at 19.2 Kbps. The PCBs are integrated into the traffic channel by robbing selected bits from the baseband stream. This way, a separate "channel" at 800 bps (for power-control purposes) exists beneath the traffic channel. The stream of PCBs at 800 bps is therefore called the *power-control subchannel* (PCS). These PCBs are continuously transmitted to the mobile by the base station. Note that since the rate of PCB transmission is 800 bps, a PCB is sent once every (1/800) second, or 1.25 ms.

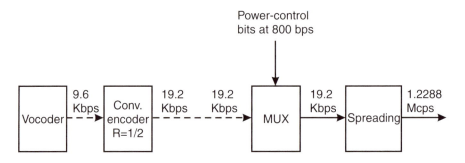

Figure 4.10 In the forward traffic channel, the PCBs at 800 bps are multiplexed directly onto the baseband information stream at 19.2 Kbps.

Both forward-link and reverse-link traffic channel frames are 20 ms in duration. Since one PCB is sent once every 1.25 ms, each traffic channel frame can be divided into (20 ms/1.25 ms) or 16 segments. These segments are called *power-control groups* (PCGs). Since each power-control group is 1.25 ms in duration and the baseband is at a rate of 19.2 Kbps, then each power-control group contains $(19.2 \times 10^{3})(1.25 \times 10^{-3}) = 24$ bits. Figure 4.11 illustrates the traffic channel frame structure.

The closed-loop power-control process is illustrated in the example shown in Figure 4.12. For example, for PCG7, the base station measures the SNR or E_b/N_0. The base station compares the measured E_b/N_0 with the threshold. If the measured E_b/N_0 is greater than the threshold, then the base station inserts a PCB of 1 during PCG9 on the forward traffic channel. If the measured E_b/N_0 is less than the threshold, then the base station inserts a PCB of 0 during PCG9 on the forward traffic channel. This process is repeated for every power-control group in the frame.

Since each PCG contains 24 bits (see Figure 4.11), the PCB can be inserted in any one of the first 16 bit positions. The exact location of the PCB in the PCG is determined in a pseudorandom fashion. The PCB bit position is determined by the decimal value of the four most significant bits of the decimator output. The input of the decimator is the long PN code. It is important to recognize that the exact location of the PCB in the PCG is not fixed but pseudorandom.

There are three additional points to mention regarding closed-loop power control.

Power-control bits are not error protected. As we can see from Figure 4.11, the PCBs are multiplexed onto the forward traffic channel *after* the convolutional encoder. Therefore, PCBs are not error protected. This is done to reduce delays that are inherent in decoding and extracting error-protected bits.

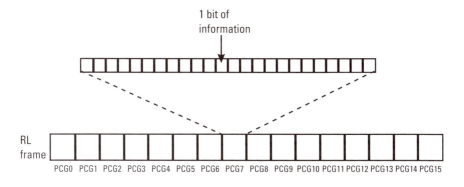

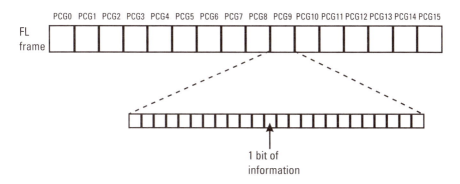

Figure 4.11 The relationship between a traffic channel frame and a PCG.

Remember that the closed-loop power control is used to combat fast Rayleigh fading; the PCBs are not error protected so that the mobile can quickly recover the PCB and adjust its transmit power accordingly. As a result, the probability of bit error for the power-control subchannel may be higher than that of the traffic channel if no special provision is taken.

Closed-loop power control has an inner and an outer loop. We have thus far only described the *inner* loop of the closed-loop power-control process. The premise of the inner loop is that there exists a predetermined SNR threshold by which power-up and power-down decisions are made. Since we are always trying to maintain an acceptable FER, and since in a mobile environment there is no one-to-one relationship between FER and E_b/N_0, the E_b/N_0 threshold has to be dynamically adjusted to maintain an acceptable FER. This adjustment of E_b/N_0 threshold (used by the inner-loop power control) is referred to as the *outer* loop of the closed-loop power control (see Figure 4.13). The outer-loop process is not defined by the IS-95 standard, and each infrastructure

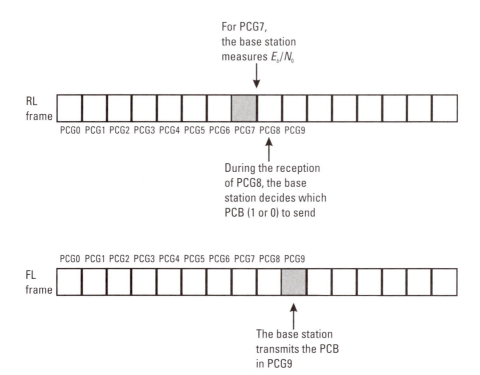

Figure 4.12 Closed-loop power control using PCBs.

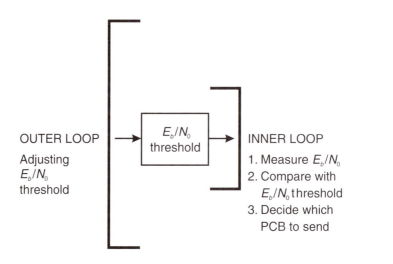

Figure 4.13 Inner and outer loops of the closed-loop power control.

manufacturer is free to implement its own outer-loop algorithms. Note that these algorithms are almost always proprietary to the manufacturer.

The final point concerns *closed-loop power control during soft handoff.* The IS-95 CDMA system utilizes *soft* handoff when a mobile moves between two or three cells. Soft handoff is the process by which a mobile maintains connection with two or three base stations as it transitions between them. During soft handoff, the mobile receives traffic channel frames from two or three base stations, and on these traffic channels there may be conflicting power-control commands (i.e., one base station may be telling the mobile to power up while the other base station may be telling the mobile to power down). In these situations, the mobile follows this rule: if any one base station commands the mobile to power down, the mobile will power down. The mobile will only power up if all of the base stations involved in soft handoff command the mobile to power up.

4.3.2.4 Open-Loop and Closed-Loop Implementation

The mobile transmit power is therefore a function of the open-loop and closed-loop power control of the system. Equation (4.16) can be modified to include the closed-loop power correction; that is,

$$
\begin{aligned}
p_t = -p_r - 73 + &\text{NOM\_PWR} + \text{INIT\_PWR} \\
+ &\text{ (sum of all access probe corrections)} \\
+ &\text{ (closed - loop correction)}
\end{aligned}
\tag{4.17}
$$

Figures 4.14(a) and 4.14(b) show one implementation of the reverse-link power-control scheme [3]. For the closed-loop power control, the base station has the entire outer loop as well as part of the inner loop; the mobile has the other part of the inner loop. For the open-loop power control, the entire open-loop portion resides in the mobile.

In Figure 4.14(a), the base station receives the reverse-link signal from the mobile. The base station first demodulates the signal and estimates the FER of the reverse link. This information on the reverse-link frame quality is fed into a threshold computer, which adjusts the E_b/N_0 threshold based on the received frame quality. At the same time, the base station also makes an E_b/N_0 estimate of the reverse link. The E_b/N_0 threshold and the E_b/N_0 estimate are then compared. If the estimate is greater than the threshold, then the link E_b/N_0 is higher than what is needed to maintain a good frame quality; a PCB of 1 is thus sent to command the mobile to power down. If the estimate is less than the threshold, then the link E_b/N_0 is lower than what is needed to maintain a good frame quality; a PCB of 0 is thus sent to command the mobile to power up.

The PCBs are multiplexed onto the forward traffic channel and transmitted to the mobile.

On the mobile side (see Figure 4.14(b)), the mobile receives the forward-link signal. It recovers the PCB and, based on the PCB, makes a decision to power up by 1 dB or to power down by 1 dB. The decision is the closed-loop correction. The correction is combined with the open-loop terms, and the combined result is fed to the transmitter so that it can transmit at the proper power level.

4.3.3   Forward Link

Ideally, power control is not needed in the forward link. The reason is that the base station is transmitting all the channels coherently in the same RF band. As Figure 4.15 shows, if thermal noise and background noise are negligible, then all the users fade together as the composite signal arrives at the mobile. However, in real life, one particular mobile may be nearby a significant jammer and experience a large background interference, or a mobile may suffer a large path loss such that the arriving composite signal is on the order of the thermal noise. Thus, forward power control is still needed. In general, however, the power-control requirement for the forward link is not as stringent as that for the reverse link.

The IS-95 standard specifies that the mobile has to report back to the base station the quality of the forward link. The mobile continuously monitors the FER of the forward link, and it reports this FER back to the base station in a message called the *power measurement report message* (PMRM). It may send this report in one of two ways: one is that the mobile periodically reports the PMRM, and the other is that the mobile reports the PMRM only if the FER exceeds a certain threshold. The base station, knowing the quality of the forward link, may then adjust its transmit power to that particular mobile. The exact algorithm of power allocation is again up to the individual infrastructure manufacturer. The process is almost always proprietary to the particular manufacturer.

4.4   Handoff

In a mobile communications environment, as a user moves from the coverage area of one base station to the coverage area of another base station, a handoff must occur to transition the communication link from one base station to the next. The CDMA system as defined by IS-95 supports different handoff processes.

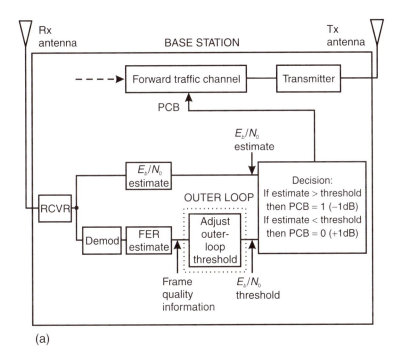

(a)

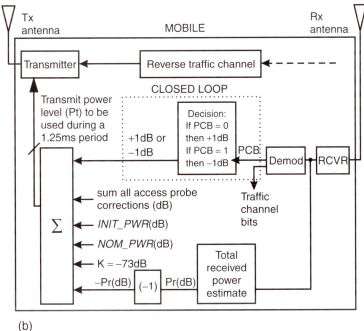

(b)

Figure 4.14 (a) Reverse-link power-control functions carried out by the base station; (b) Reverse-link power-control functions carried out by the mobile. After [3].

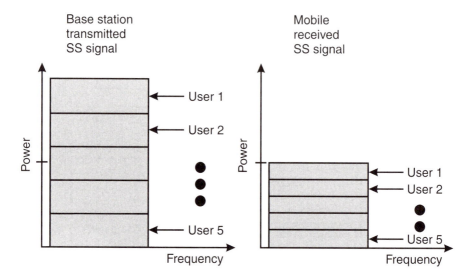

Figure 4.15 All users fade together as the composite spread-spectrum signal travels from the base station to the mobile.

The first is the *soft handoff*. In Section 4.3.2, we briefly mentioned that CDMA uses soft handoff where, during handoff, a mobile simultaneously maintains connection with two or three base stations. As the mobile moves from its current cell (source cell) to the next cell (target cell), a traffic channel connection is simultaneously maintained with both cells. Figure 4.16(a) and Figure 4.16(b) illustrate the simultaneous links during soft handoff. On the forward link (see Figure 4.16(a)), the mobile uses the rake receiver to demodulate two separate signals from two different base stations. The two signals are combined to yield a composite signal of better quality. On the reverse link (see Figure 4.16(b)), the mobile's transmit signal is received by both base stations. The two cells demodulate the signal separately and send the demodulated frames back to the *mobile switching center* (MSC). The MSC contains a selector that selects the best frame out of the two that are sent back.

The second is the *softer handoff*. This type of handoff occurs when a mobile transitions between two different sectors of the same cell. On the forward link, the mobile performs the same kind of combining process as that of soft handoff. In this case, the mobile uses its rake receiver to combine signals received from two different sectors. On the reverse link, however, two sectors of the same cell simultaneously receive two signals from the mobile. The signals are demodulated and combined inside the cell, and only one frame is sent back to the MSC.

FORWARD LINK

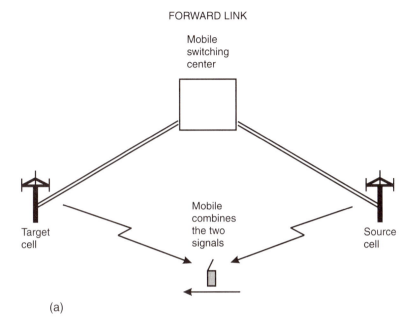

(a)

REVERSE LINK

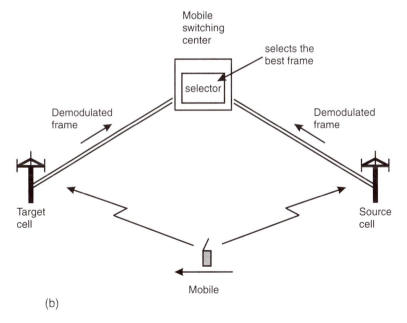

(b)

Figure 4.16 (a) Soft handoff between two base stations—forward link; (b) soft handoff between two base stations—reverse link.

The third is the *hard handoff.* The CDMA system uses two types of hard handoffs. *CDMA-to-CDMA handoff* occurs when the mobile is transitioning between two CDMA carriers (i.e., two spread-spectrum channels that are centered at different frequencies). This hard handoff can also occur when the mobile is transitioning between two different operators' systems. CDMA-to-CDMA handoff is sometimes called *D-to-D handoff.* On the other hand, *CDMA-to-analog handoff* occurs when a CDMA call is handed down to an analog network. This can occur when the mobile is traveling into an area where there is analog service but no CDMA service. CDMA-to-analog handoff is sometimes called *D-to-A handoff.*

Before we describe the soft handoff processes in detail, it is important to note that each sector in a CDMA system is distinguished from one another by the pilot channel of that sector. As Figure 4.17 shows, the pilot channel is one of the four logical channels—pilot, paging, sync, and traffic channels—on the forward link. The pilot channel serves as a "beacon" for the sector and aids the mobile in acquiring other logical channels of the same sector. There is no information contained in the pilot other than the short PN code with a specific *offset* assigned to that particular sector. Remember from Chapter 3 that a PN sequence with an offset becomes another PN sequence, and this offset PN sequence is orthogonal to the original sequence. The PN code transmitted on the pilot channel uses this quality to distinguish itself from other sectors and

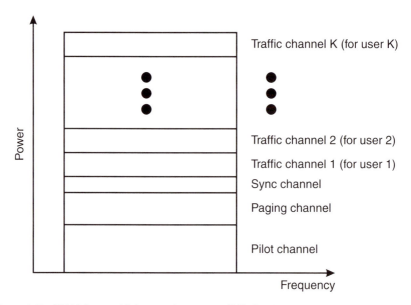

Figure 4.17 CDMA forward link spread spectrum (SS) signal.

other base stations. The offset of a PN sequence associated with a particular sector is specified in the parameter PILOT_PN for that sector.

A special term is used to describe the SNR of the pilot channel: energy per chip per interference density, or E_c/I_0. The energy per chip E_c is different from energy per bit E_b in that "chips" refer to PN sequences that are spread. Since there is no baseband information contained in the pilot channel, the pilot is not despread and bits are not recovered. Therefore, in order to describe the signal strength of the pilot channel, the raw SNR, or E_c/I_0, is used. Note that since the pilot is not despread, E_c/I_0 remains less than 1 most of the time.

4.4.1 Set Maintenance

In CDMA, the mobile is an intimate participant in the soft handoff process. The mobile constantly notifies the base station regarding the local propagation condition; the base station makes use of this information to make handoff decisions. This *mobile-assisted handoff* (MAHO) is evident in that the mobile makes a measurement of forward link E_c/I_0 and reports the measurement result to the base station. Since each base station transmits its own pilot on a different PN offset, the E_c/I_0 of a pilot gives a good indication of whether or not the particular sector should be the serving sector for the mobile.

In managing the handoff process, the mobile maintains in its memory four exclusive lists of base station sectors. The sectors are stored in the form of pilot PN offsets of those sectors. These lists are also called *sets*. The four sets are *active* set, *candidate* set, *neighbor* set, and *remaining* set [2].

The active (A) set contains the pilots of those sectors that are actively communicating with the mobile on traffic channels. If the active set contains only one pilot, then the mobile is not in soft handoff. If the active set contains more than one pilot, then the mobile is maintaining connection with all those sectors on separate traffic channels. The base station ultimately controls the handoff process because a pilot can only be added to the active set if the base station sends a *handoff direction message* to the mobile and the message contains that particular pilot to be added to the active set. The active set can contain at least six pilots.

The candidate set contains those pilots whose E_c/I_0s are sufficient to make them handoff candidates. This means that if the E_c/I_0 of a particular pilot is greater than the *pilot detection threshold* T_ADD, then that pilot will be added to the candidate set. A pilot is removed from this set and placed in the neighbor set if the strength of that pilot drops below the *pilot drop threshold* T_DROP for more than the duration specified by the *handoff drop timer expiration value* T_TDROP. The candidate set can contain at least six pilots.

Note that a pilot can be removed from the active set and placed in the candidate set if the received *handoff direction message* does not include that particular pilot; and if T_TDROP for that pilot has not expired, the pilot is removed from the active set and placed in the candidate set.

The neighbor (N) set contains those pilots that are in the *neighbor list* of the mobile's current serving sector. Initially, the neighbor set contains those pilots that are sent to the mobile in the *neighbor list message* by the serving base station. In order to keep current all the pilots in the neighbor set, the mobile keeps an aging counter for each pilot in this set. The counter is initialized to zero when the pilot is moved from the active or candidate set to the neighbor set. The counter is incremented for each pilot in the neighbor set whenever a *neighbor list update message* is received. The pilot is moved from this set to the remaining set if the counter exceeds NGHBR_MAX_AGE. The neighbor set can contain at least 20 pilots.

Note that a pilot can be removed from the active set and placed in the neighbor set if the received *handoff direction message* does not include that particular pilot; and if T_TDROP for that pilot has expired, the pilot is removed from the active set and placed in the neighbor set.

The remaining (R) set contains all possible pilots in the system for this CDMA carrier frequency, excluding pilots that are in active, candidate, and neighbor sets. The pilot PN offsets in this set are defined by the parameter pilot increment PILOT_INC. For example, if PILOT_INC is 4, then individual sectors in the system can only transmit pilots with offsets of 0, 4, 8, 12, and so forth. PILOT_INC is sent to the mobile in the *neighbor list message* and *neighbor list update message.*

4.4.2 Handoff Process

In the following example, we examine the handoff process from the source cell to the target cell. As Figure 4.18 shows, the mobile is moving from the coverage area of source cell A to the coverage area of target cell B. The following is a sequence of events during this transition:

1. The mobile here is being served by cell A only, and its active set contains only pilot A. The mobile measures pilot B E_c/I_0 and finds it to be greater than T_ADD. The mobile sends a *pilot strength measurement message* and moves pilot B from the neighbor set to the candidate set.

2. The mobile receives a *handoff direction message* from cell A. The message directs the mobile to start communicating on a new traffic

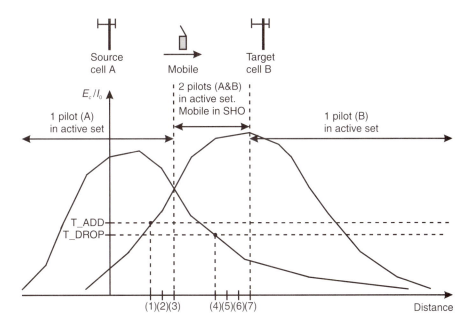

Figure 4.18 The handoff process. After [1].

channel with cell B; the message contains the PN offset of cell B and the Walsh code of the newly assigned traffic channel.

3. The mobile moves pilot B from the candidate set to the active set. After acquiring the forward traffic channel specified in the *handoff direction message*, the mobile sends a *handoff completion message*. Now the active set contains two pilots.

4. The mobile detects that pilot A has now dropped below T_DROP. The mobile starts the drop timer.

5. The drop timer reaches T_TDROP. The mobile sends a *pilot strength measurement message*.

6. The mobile receives a *handoff direction message*. The message contains only the PN offset of cell B. The PN offset of cell A is not included in the message.

7. The mobile moves pilot A from the active set to the neighbor set, and it sends a *handoff completion message*.

There is another mechanism that can trigger the transmission of a *pilot strength measurement message* by the mobile. If the strength of a pilot in the candidate set exceeds the strength of a pilot in the active set by the *active set versus* candidate set comparison threshold T_COMP × 0.5 dB, then the mobile sends a pilot strength measurement message.

4.4.3 Pilot Search

In addition to being spread by the Walsh code, the forward link is also spread by a PN sequence (with a specific PILOT_PN offset) at 1.2288 Mcps. This forward-link signal, like any other signal traveling through a mobile environment, can encounter reflections that result in multipaths. As a result, different pilot signals can arrive at the mobile at different times, and a multipath component of one pilot may arrive a few chips later than its direct-path component. Therefore, search windows are provided to search for pilots that are in the active, candidate, neighbor, and remaining windows. Specifically, the parameter SRCH_WIN_A defines the search-window width used to search for pilots in the active and candidate sets, the parameter SRCH_WIN_N defines the search-window width used to search for pilots in the neighbor sets, and the parameter SRCH_WIN_R defines the search-window width used to search for pilots in the remaining sets. These three parameters are sent to the mobile in the *system parameters message* and *handoff direction message*.

The search window for the active and candidate sets is referenced to the earliest arriving multipath component of the pilot. The mobile should center the search window for each pilot in the active and candidate sets around the earliest arriving usable multipath component of the pilot. For example, if SRCH_WIN_A is defined to be 40 chips, then the mobile searches 20 chips around the earliest arriving multipath component of the pilot. For each pilot in the neighbor and remaining sets, the mobile centers the search window for each pilot around the pilot's PN sequence offset using the mobile's timing reference [1].

References

[1] TIA/EIA IS-95A, "Mobile Station-Base Station Compatibility Standard for Dual-Mode Wideband Spread Spectrum Cellular System," Telecommunications Industry Association.

[2] Qualcomm, *The CDMA Network Engineering Handbook, Vol. 1: Concepts in CDMA*, 1993.

[3] Labedz, G., et al., "Predicting Real-World Performance for Key Parameters in a CDMA Cellular System," *Proc. 46th Annual Vehicular Technology Conference*, IEEE, 1996, pp.1472–1476.

Select Bibliography

Feher, K., *Wireless Digital Communications Modulation and Spread Spectrum Applications*, Englewood Cliffs, NJ: Prentice Hall, 1995.

Gilhousen, K. S., et al., "On the Capacity of a Cellular CDMA System," *IEEE Trans. on Vehicular Technology*, Vol. 40, May 1991, pp. 306–307.

Glisic, S., and B. Vucetic, *Spread Spectrum CDMA Systems for Wireless Communications*, Norwood, MA: Artech House, 1997.

Harte, L., *CDMA IS-95 for Cellular and PCS: Technology, Applications and Resource Guide*, New York, NY: McGraw-Hill, 1997.

Kohno, R., R. Meidan, and L. B. Milstein, "Spread Spectrum Access Methods for Wireless Communications," *IEEE Communications Magazine*, Jan. 1995.

Padovani, R. "Reverse Link Performance of IS-95 Based Cellular Systems," *IEEE Personal Communications Magazine*, Third Quarter, 1994.

Peterson, R. L., R. E. Ziemer, and D. E. Borth, *Introduction to Spread-Spectrum Communications*, Englewood Cliffs, NJ: Prentice Hall, 1995.

Pickholtz, R. L., D. L. Schilling, and L. B. Milstein, "Theory of Spread-Spectrum Communications—A Tutorial," *IEEE Trans. on Communications,* Vol. COM-30, No. 5, May 1982, p. 855.

Viterbi, A. J., et al., "Soft Handoff Extends CDMA Cell Coverage and Increases Reverse Link Capacity," *IEEE J. in Selected Areas in Communications*, Vol. 12, No. 8, 1994.

Viterbi, A. J., *CDMA Principles of Spread Spectrum Communication*, New York, NY: Addison-Wesley, 1995.

Viterbi, A. J., A. M. Viterbi, and E. Zehavi, "Performance of Power-Controlled Wideband Terrestrial Digital Communication," *IEEE Trans. on Communications*, Vol. 41, No. 4, 1993.

Viterbi, A. J., and R. Padovani, "Implications of Mobile Cellular CDMA," *IEEE Communications Magazine*, Dec. 1992.

5

Link Structure

5.1 Asymmetric Links

The IS-95 CDMA system is unique in that its forward and reverse links have different link structures. This is necessary to accommodate the requirements of a land-mobile communication system. The forward link consists of four types of logical channels: pilot, sync, paging, and traffic channels. There is one pilot channel, one sync channel, up to seven paging channels, and several traffic channels. Each of these forward-link channels is first spread orthogonally by its Walsh function, then it is spread by a quadrature pair of short PN sequences. All channels are added together to form the composite SS signal to be transmitted on the forward link.

The reverse link consists of two types of logical channels: access and traffic channels. Each of these reverse-link channels is spread orthogonally by a unique long PN sequence; hence, each channel is identified using the distinct long PN code. The reason that a pilot channel is not used on the reverse link is that it is impractical for each mobile to broadcast its own pilot sequence.

5.2 Forward Link

In Section 3.5.1, we defined the structure of a Hadamard matrix and described how Walsh codes are generated using such a matrix. The IS-95 CDMA system uses a 64 by 64 Hadamard matrix to generate 64 Walsh functions that are orthogonal to each other, and each of the logic channels on the forward link is identified by its assigned Walsh function.

5.2.1 Pilot Channel

The pilot channel is identified by the Walsh function 0 (w_0). The channel itself contains no baseband information. The baseband sequence is a stream of 0s that are spread by Walsh function 0, which is also a sequence of all 0s. The resulting sequence (still all 0s) is then spread, or multiplied, by a pair of quadrature PN sequences. Therefore, the pilot channel is effectively the PN sequence itself (see Figure 5.1). The PN sequence with a specified offset uniquely identifies the particular sector that is transmitting the pilot signal. Note that both Walsh function 0 and the PN sequence are running at a rate of 1.2288 Mcps. After PN spreading, baseband filters are used to shape the digital pulses. These filters effectively lowpass filter the digital pulse stream and control the baseband spectrum of the signal. This way, the signal bandwidth may have a sharper roll-off near the band edge.

The pilot channel is transmitted continuously by the base station sector. The pilot channel provides the mobile with timing and phase reference. The mobile's measurement of the signal-to-noise ratio (i.e., E_c/I_0) of the pilot channel also gives an indication of which is the strongest serving sector of that mobile.

5.2.2 Sync Channel

Unlike the pilot channel, the sync channel carries baseband information. The information is contained in the *sync channel message* that notifies the mobile

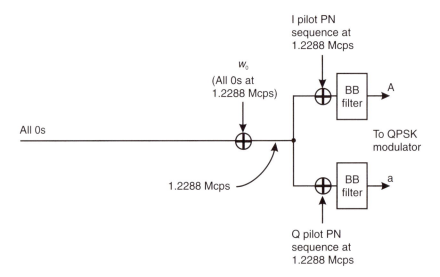

Figure 5.1 Pilot channel. After [1].

of important information about system synchronization and parameters. Figure 5.2 shows that the baseband information is error protected and interleaved. It is then spread by Walsh function 32 and further spread by the PN sequence that is identified with the serving sector. The baseband information is at a rate of 1.2 Kbps.

At the bit level, the sync channel is transmitted in groups of *sync channel superframes*, each superframe contains 96 bits and lasts 90 ms, yielding a data rate of (96 bits/90 ms) = 1,200 bps. Each superframe contains three *sync channel frames* of equal length and duration (see Figure 5.3). Each sync channel frame is aligned with the short PN sequence associated with the transmitting sector. Note that the short PN sequence repeats every 26.67 ms, and *each period of the short PN sequence is synchronized with each sync channel frame.* Therefore, once the mobile acquires synchronization with the pilot channel, the alignment for the sync channel is immediately known. This is because the sync channel is spread with the same pilot PN sequence, and because the frame timing of the sync channel is aligned with that of the pilot PN sequence [1]. Once the mobile achieves alignment with the sync channel, the mobile can start reading the *sync channel message*.

The *sync channel message* itself is long and may occupy more than one sync channel frame. Therefore, the *sync channel message* is organized in a structure called the *sync channel message capsule*. A sync channel message capsule consists of the *sync channel message* and *padding*. The sync channel message resides in more than one sync channel frame, and padding (of bits) is used to fill up the bit positions all the way up to the beginning of the next sync channel superframe, where the next *sync channel message* starts.

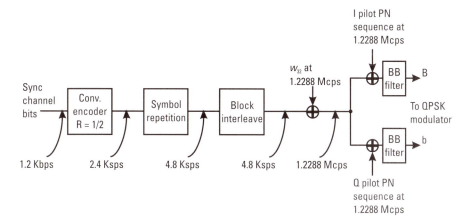

Figure 5.2 Sync channel. After [1].

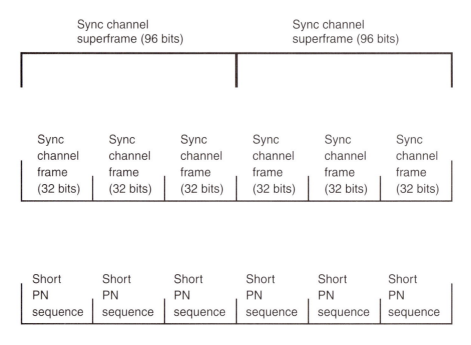

Figure 5.3 Sync channel frame structure.

Each sync channel frame begins with the *start-of-message* (SOM) bit. The SOM bit is the first bit of the sync channel frame; an SOM of 1 indicates the start of the *sync channel message*, and an SOM of 0 indicates that the current sync channel frame has the contents of a running *sync channel message* that started in some previous frame. This way, the base station can transmit the sync channel message in consecutive sync channel frames. Note that an SOM of 1 also coincides with the start of a sync channel superframe. In other words, a sync channel message always starts at the beginning of a sync channel superframe. Note that each sync channel frame starts with the SOM bit, and the rest of the frame is referred to as the *sync channel frame body*. Figure 5.4 shows the structure of a hypothetical *sync channel message* that occupies two consecutive superframes.

The *sync channel message* itself contains different fields; the message contains information such as the offset of the pilot PN sequence used by the transmitting sector (i.e., the PILOT_PN field). The message also contains information to enable the mobile to synchronize with the long PN sequence. This is done by reading the LC_STATE and SYS_TIME fields of the *sync channel message*. The base station sets the LC_STATE field to the long-code state at some future time given by the SYS_TIME field. And at the precise time given

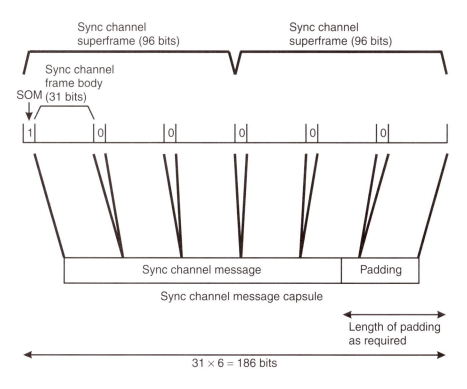

Figure 5.4 Sync channel message structure for a hypothetical message that occupies two consecutive superframes. After [1].

by the SYS_TIME field, the mobile starts running the long PN sequence (at the state given by the LC_STATE field). Thus, after the successful reception of the *sync channel message*, the mobile is synchronized with the short PN sequence transmitted by the base station, the mobile knows the exact PN offset of that short PN sequence, and the mobile is synchronized with the long PN sequence.

5.2.3 Paging Channel

Similar to the sync channel, the paging channel also carries baseband information. But unlike the sync channel, the paging channel transmits at higher rates; it can transmit at either 4.8 or 9.6 Kbps. The PRAT field in the sync channel message informs the mobile of the data rate of the paging channel. Once the mobile acquires timing and synchronization using the sync channel, the mobile begins to monitor the paging channel. Although there can be up to seven paging channels per sector, each mobile only monitors one paging channel.

As shown in Figure 5.5, the baseband information is first error protected, then if the data rate is at 4.8 Kbps, the bits are repeated once. Otherwise, they are not repeated. Following interleaving, the data is first scrambled by a decimated long PN sequence, then it is spread by a specific Walsh function assigned to that paging channel and further spread by the short PN sequence assigned to the serving sector. Also note from Figure 5.5 that the long PN code undergoes a decimation ratio of 64:1 (i.e., from 1.2288 Mcps to 19.2 Ksps). The long-code generator itself is masked with a mask specific to each unique paging channel number (i.e., 1 through 7). Therefore, the long-code mask used for paging channel 1 (spread by Walsh function 1) is different from that used for paging channel 3 (spread by Walsh function 3).

The paging channel is divided into 80-ms slots. A group of 2,048 slots is called a *maximum slot cycle*. An 80-ms slot is divided into four paging channel *frames*, and each paging channel frame is further divided into two paging channel *half-frames*. The first bit of each half-frame is called the *synchronized capsule indicator* (SCI) bit. Figure 5.6 depicts the frame structure of paging channel.

A message on the paging channel may occupy more than one paging channel half-frames, and a message may end in the middle of a paging channel half-frame. The message on the paging channel may be transported by either

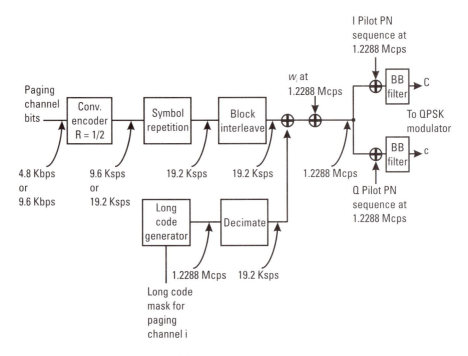

Figure 5.5 Paging channel. After [1].

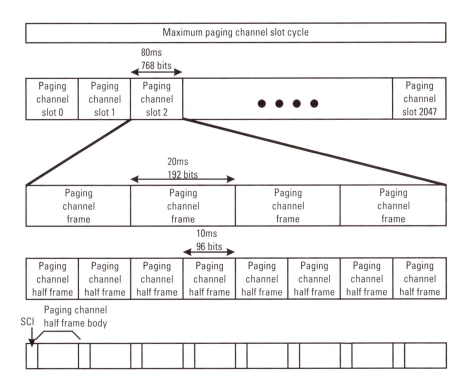

Figure 5.6 Paging channel frame structure for a paging channel rate of 9.6 Kbps. After [1].

synchronized paging channel message capsules or *unsynchronized paging channel message capsules.* If a message ends in the middle of a paging channel half-frame, and if there are less than eight bits between the end of that message and the SCI bit of the next half-frame, then the base station would include enough padding bits at the end of the current message capsule to extend the capsule up to the beginning of the next half-frame. If the next message capsule to be transmitted is a synchronized message capsule, then the base station would also include enough padding bits at the end of the current message capsule to extend the capsule up to the beginning of the next half-frame.

However, if a message ends in the middle of a paging channel half-frame, and if there are more than eight bits between the end of that message and the SCI bit of the next half-frame, then the base station may transmit an unsynchronized message capsule immediately after that message. In this case, no padding bits are added.

Therefore, the SCI bit flags the start of a brand new message capsule in the current half-frame (i.e., if the SCI bit is 1). Then a new message capsule starts immediately follows that SCI bit. The SCI bit is set to 0 in all other cases.

Figure 5.7 shows three hypothetical paging channel messages that are transmitted consecutively.

Messages such as *overhead* and *paging* are sent over the paging channel. Overhead information is used to notify the mobile of important system configuration parameters. Examples of overhead messages are the *system parameters message*, the *access parameters message*, and the *neighbor list message*. The *system parameters message* contains important system configuration parameters; these include the following:

- Handoff parameters for the mobile to use [1]:

    T_ADD—pilot detection threshold;

    T_DROP—pilot drop threshold;

    T_COMP—active set versus candidate set comparison threshold;

    T_TDROP—drop timer value;

    SRCH_WIN_A—search-window size for the active and candidate sets;

    SRCH_WIN_N—search-window size for the neighbor set;

    SRCH_WIN_R—search-window size for the remaining set;

    NGHBR_MAX_AGE—neighbor set maximum age.

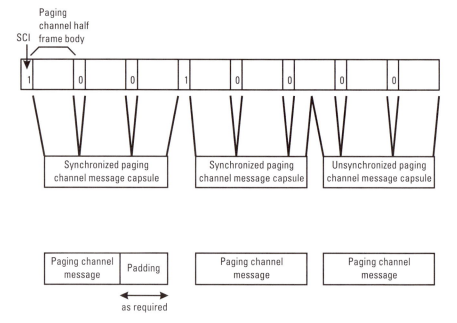

Figure 5.7 Paging channel message structure for three different hypothetical messages, each of which occupies more than two paging channel half-frames. After [1].

- Forward power-control parameters for the mobile to use [1]:
  PWR_REP_THRESH—power-control reporting threshold;
  PWR_REP_FRAMES—power-control reporting frame count;
  PWR_THRESH_ENABLE—threshold report mode indicator;
  PWR_PERIOD_ENABLE—periodic report mode indicator;
  PWR_REP_DELAY—power report delay.

The *access parameters message* contains important access configuration parameters; these include the following:

- Reverse power-control parameters for the mobile to use [1]:
  NOM_PWR—nominal transmit power offset;
  INIT_PWR—initial power offset for access;
  PWR_STEP—power increment.

- Access parameters for the mobile to use [1]:
  NUM_STEP—number of access probes;
  PROBE_PN_RAN—time randomization for access channel probes;
  ACC_TMO—acknowledgment timeout;
  PROBE_BKOFF—access channel probe backoff range;
  BKOFF —access channel probe sequence backoff range;
  MAX_REQ_SEQ—maximum number of access probe sequences for an access channel request;
  MAX_RSP_SEQ—maximum number of access probe sequences for an access channel response.

- Access channel parameters for the mobile to use [1]:
  MAX_CAP_SZ—maximum access channel message capsule size;
  PAM_SZ—access channel preamble length.

The *neighbor list message* contains a list of neighboring sectors for the mobile to use; these include PN offsets of neighbors (i.e., NGHBR_PN) [1].

In addition, paging messages are sent over the paging channel. A *page message* may contain a page to one particular mobile, or it may contain a page to a group of mobiles.

5.2.4   Traffic Channel

The forward traffic channel is used to transmit user data and voice; signaling messages are also sent over the traffic channel. The structure of the forward traffic channel is similar to that of the paging channel. The only difference is that the forward traffic channel contains multiplexed PCBs, which are discussed in Chapter 4.

Figure 5.8 shows the forward traffic channel for Rate Set 1. For this rate set, the vocoder is capable of varying its output data rate in response to speech activities. Four different data rates are supported: 9.6, 4.8, 2.4, and 1.2 Kbps. For example, during quiet periods of speech, the vocoder may elect to code the speech at the lowest rate of 1.2 Kbps.

The baseband data from the vocoder is convolutionally encoded for error protection. For Rate Set 1, a rate 1/2 convolutional encoder is used. The encoding effectively doubles the data rate. After convolutional encoding, the data undergoes symbol repetition, which repeats the symbols when lower rate data are produced by the vocoder. The following is the repetition scheme:

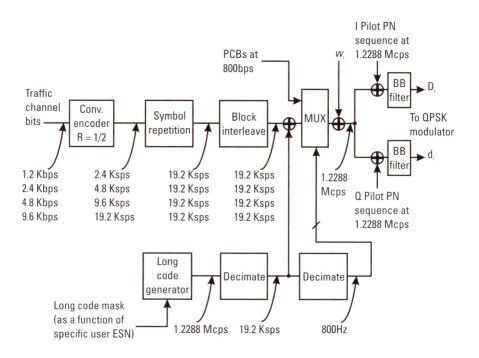

Figure 5.8 Forward traffic channel for Rate Set 1. After [1].

- When the data rate is 9.6 Kbps, the code symbol rate (at the output of the convolutional encoder) is 19.2 Ksps. In this case, no repetition is performed.

- When the data rate is 4.8 Kbps, the code symbol rate is 9.6 Ksps; each symbol is repeated once, yielding a final modulation symbol rate of 19.2 Ksps.

- When the data rate is 2.4 Kbps, the code symbol rate is 4.8 Ksps; each symbol is repeated three times, yielding a final modulation symbol rate of 19.2 Ksps.

- When the data rate is 1.2 Kbps, the code symbol rate is 2.4 Ksps; each symbol is repeated seven times, yielding a final modulation symbol rate of 19.2 Ksps.

The reason for repeating symbols is to reduce overall interference power at a given time when lower rate data are transmitted. Figures 5.9(a) and 5.9(b) illustrate the concept. Two cases are presented in Figures 5.9(a) and 5.9(b). In the first case, the system wants to transmit at a rate of 2 symbols per second, and in the second case, the system wants to transmit at a lower rate of 1 symbol per second. We further assume that the integrator needs to accumulate 1W of signal power in order to successfully demodulate a symbol.

In the first case, the original symbols are not repeated. If each symbol is sent at an energy level of 2, then the integrator (in the receiver) is able to accumulate 1W during each original symbol period. The symbols in this case are successfully demodulated. Note that in this case, the original symbol period is 0.5 second.

In the second case, the original symbols are repeated once, yielding a symbol rate of 2 symbols per second. However, since the original symbol rate is only 1 symbol per second, we can lower the energy of each *repeated* symbol by one-half. This way, when the integrator accumulates energy over the *original* symbol period (i.e., 1 sec), the integrator would still accumulate the necessary 1W per each original symbol period. Note that in this case, the original symbol period is 1 sec. The motivation for the symbol repetition scheme is to *decrease the power per repeated (transmitted) symbol* when the vocoder is running at a lower rate. The scheme is effectively a way of taking advantage of the voice activity factor in hardware implementation (i.e., when the vocoder is transmitting at lower rates, the forward transmit power is reduced).

In a real CDMA system, when the vocoder is transmitting at 4.8 Kbps, the energy per symbol transmitted is one-half that of 9.6 Kbps. When the

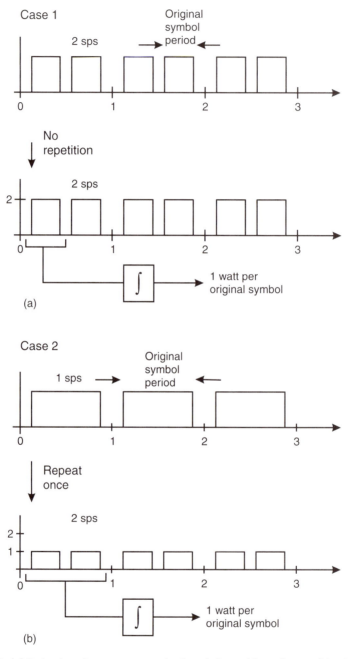

Figure 5.9 (a) Reduction of power as a result of symbol repetition—Case 1; (b) reduction of power as a result of symbol repetition—Case 2.

vocoder is transmitting at 2.4 Kbps, the energy per symbol transmitted is one-forth that of 9.6 Kbps, and when the vocoder is transmitting at 1.2 Kbps, the energy per symbol transmitted is one-eighth that of 9.6 Kbps.

After symbol repetition, the data is interleaved to combat fading (see Figure 5.8), then the interleaved data is scrambled by a decimated long PN sequence. The long PN sequence is generated by a long PN code generator. The generator outputs a long PN sequence at 1.2288 Mcps. Because the data rate at the interleaver output is 19.2 Ksps, the PN sequence is *decimated* by a ratio of 64:1 to also achieve a rate of 19.2 Kcps; the decimated long PN sequence at 19.2 Kcps is then multiplied with the 19.2-Ksps data stream. Note that the long-code generator produces the long PN sequence using a mask that is specific to the mobile. In reality, the mask is a function of the mobile's *electronic serial number* (ESN).

The PCBs at 800 bps are then multiplexed with the scrambled data stream at 19.2 Ksps. A PCB can be punctured into any one of the first 16 bit positions of a PCG (which contains 24 bits). The exact location of the PCB in the PCG is determined in a pseudorandom fashion. More specifically, given that the input of the decimator is the long PN sequence, the PCB bit position is determined by the decimal value of the four most significant bits of the decimator output. It is important to recognize that the exact location of the PCB in the PCG is not fixed, but is determined in a pseudorandom manner. For more details on the power control subchannel and PCB puncturing, consult Section 7.1.3.1.7 of [1].

At this point, the multiplexed data stream (still at 19.2 Ksps) is orthogonally spread by the assigned Walsh function. Each forward traffic channel is identified by its assigned Walsh function. The spreading Walsh function is at a rate of 1.2288 Mcps; each symbol is spread by a factor of 64, and the result is a spread data stream at a rate of 1.2288 Mcps.

The data stream is further spread by the assigned short PN sequence of the transmitting sector. The short PN sequence provides a second layer of isolation that distinguishes among the different transmitting sectors. This way, all 64 available Walsh functions can be reused in every sector. Remember that each unique short PN sequence is characterized by its PN offsets.

The forward traffic channel structure is similar for Rate Set 2. The Rate Set 2 vocoder codes speech at higher rates, and it delivers a better voice quality than that of Rate Set 1. The Rate Set 2 vocoder supports four variable rates: 14.4, 7.2, 3.6, and 1.8 Kbps. Figure 5.10 shows the forward traffic channel for Rate Set 2. Note that in order to maintain the output of the block interleaver at 19.2 Ksps, the rate of the convolutional encoder is increased to $R = 3/4$.

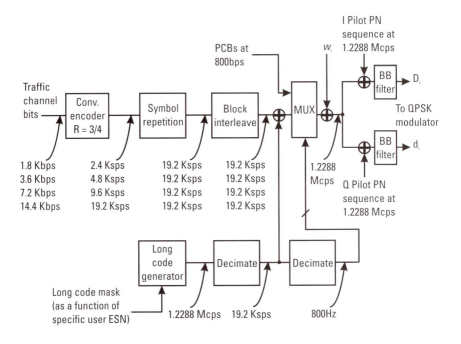

Figure 5.10 Forward traffic channel for Rate Set 2.

5.2.5 Modulator

The output of the logical channels is fed into the modulator. Figure 5.11 shows the structure of the forward channel modulator [2]. The gain of each logical channel, including pilot, sync, paging, and all traffic channels, is first adjusted by the gain control function. The gain of each channel dictates how much power is to be transmitted for that channel. The gains for the individual traffic channels are dynamically changing (i.e., they are controlled by the forward power-control process described in Chapter 4).

After the channel gains are adjusted, the signals are coherently added together to form the composite spread-spectrum signal. After the summation, both the *I* and the *Q* paths are up-converted by their respective carriers. The up-converted signals then are added together to form the final passband QPSK signal.

5.3 Reverse Link

The reverse link supports two types of logical channels: access channels and traffic channels. Because of the noncoherent nature of the reverse link, Walsh

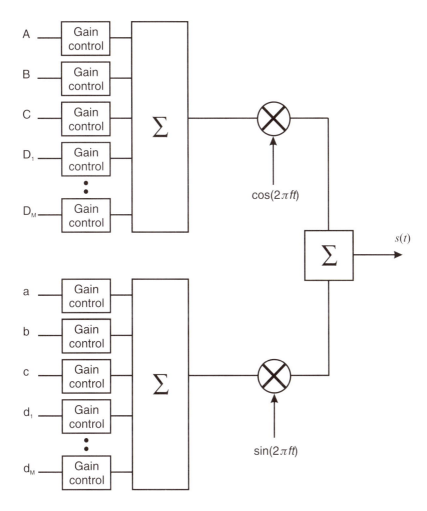

Figure 5.11 Modulator for the CDMA forward link. After [2]. Note that the inputs A, B, C, D, and a, b, c, and d are drawn from Figures 5.1, 5.2, 5.5, and 5.8.

functions are not used for channelization. Instead, long PN sequences are used to distinguish the users from one another.

5.3.1 Access Channel

The access channel is used by the mobile to communicate with the base station when the mobile doesn't have a traffic channel assigned. The mobile uses this channel to make call originations and respond to pages and orders. The base-band data rate of the access channel is fixed at 4.8 Kbps.

As shown in Figure 5.12, the baseband information is first error protected by an $R = 1/3$ convolutional encoder. The lower encoding rate makes error protection more robust on the reverse link, which is often the weaker of the two links. The symbol repetition function repeats the symbol once, yielding a code symbol rate of 28.8 Ksps. The data is then interleaved to combat fading.

Following interleaving, the data is coded by a 64-ary orthogonal modulator. The set of 64 Walsh functions is used, but here the Walsh functions are used to modulate, or represent, groups of six symbols. The reason for orthogonal modulation of the symbols is again due to the noncoherent nature of reverse link. When a user's transmission is not coherent, the receiver (at the base station) still has to detect each symbol correctly. Making a decision of whether or not a symbol is +1 or −1 may be difficult during one symbol period.

However, if a group of six symbols is represented by a unique Walsh function, then the base station can easily detect six symbols at a time by deciding which Walsh function is sent during that period. The receiver can easily decide which Walsh function is sent by correlating the received sequence with the set of 64 known Walsh functions. Note that on the forward link, Walsh functions are used to distinguish among the different channels. On the reverse link, Walsh functions are used to distinguish among the different symbols (or among groups of six symbols).

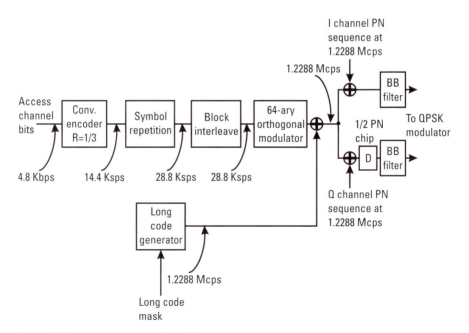

Figure 5.12 Access channel. After [1].

In reality, a group of six binary symbols corresponds to a decimal value between 0 and 63. The pattern of the six-symbol group (and the corresponding decimal value) dictates which Walsh function (0 to 63) is used to represent that group of six symbols. For example, a group of six symbols $(-1, +1, -1, +1, +1, -1)$ corresponds to a binary value of 010110, or a decimal value of 22. Thus Walsh function 22 is the output of the orthogonal modulator.

The orthogonally modulated data at 4.8 Ksps (modulation symbols) or at 307.2 Ksps (code symbols) are then spread by the long PN sequence. The long PN sequence is running at 1.2288 Mcps, and the bandwidth of the data after spreading is 1.2288 Mcps. Remember that the long PN sequence is used to distinguish the access channel from all other channels that occupy the reverse link.

The data is further scrambled in the I and the Q paths by the short PN sequences (also running at 1.2288 Mcps) defined in the IS-95 standard. Because the reverse link uses OQPSK modulation, the data in the Q path is delayed by one-half a PN chip. See Chapter 3 for a description of OQPSK. The primary purpose of this chip delay is to make sure that the QPSK signal envelope will not collapse to zero. This property is important because the power amplifier of the mobile is typically small and limited in performance. If we can ensure that the signal envelope never reaches zero and always stays above a certain level, then the amplifier would only have to remain linear over a smaller dynamic range.

Information is transmitted on the access channel in *access channel slots* and *access channel frames*. Each slot contains $(3 + \text{MAX\_CAP\_SZ}) + (1 + \text{PAM\_SZ})$ frames [1], and each frame contains 96 bits and lasts 20 ms, which corresponds to a baseband data rate of 4.8 Kbps. Recall that MAX_CAP_SZ is the maximum access channel message capsule size and PAM_SZ is the access channel preamble length. Figure 5.13 depicts the frame structure of the access channel.

As shown in Figure 5.13, although the base station allows each slot to contain $(3 + \text{MAX\_CAP\_SZ}) + (1 + \text{PAM\_SZ})$ frames, the mobile may not need that many frames in the slot to transmit its message. The mobile would set a variable CAP_SZ according to the length of the actual message. The constraint is that the CAP_SZ has to be less than $(3 + \text{MAX\_CAP\_SZ})$. Figure 5.14 shows the message structure of the access channel. We have the access channel preamble consisting of $(1 + \text{PAM\_SZ})$ frames. Each access channel frame consists of 88 information bits and 8 encoder tail bits. For those frames that are not in the preamble, the information bits in the frame bodies are combined to form the *access channel message capsule,* which contains $(\text{CAP\_SZ} \times 88)$ bits. The access channel message capsule contains the *access channel message* and *padding.* Again, the *access channel message* may occupy more

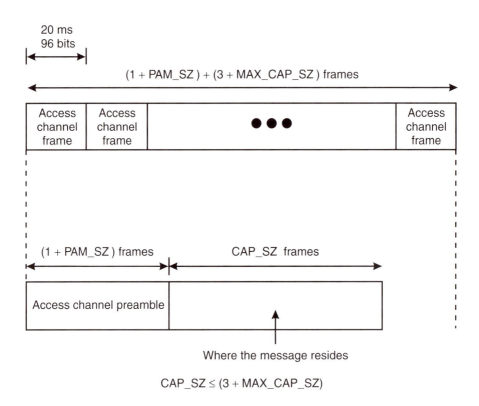

CAP_SZ ≤ (3 + MAX_CAP_SZ)

Figure 5.13 Access channel frame structure.

than one frame, and padding bits are added so that the total length of the *access channel message* and padding is equal to (CAP_SZ × 88) bits.

There are two types of messages sent on the access channel: a *response* message (in response to a base station message) or a *request* message (sent by the mobile station). Different procedures are used to send these two types of messages. Chapter 6 describes these procedures in more detail.

5.3.2 Traffic Channel

The reverse traffic channel is used to transmit user data and voice; signaling messages are also sent over the traffic channel. The structure of the reverse

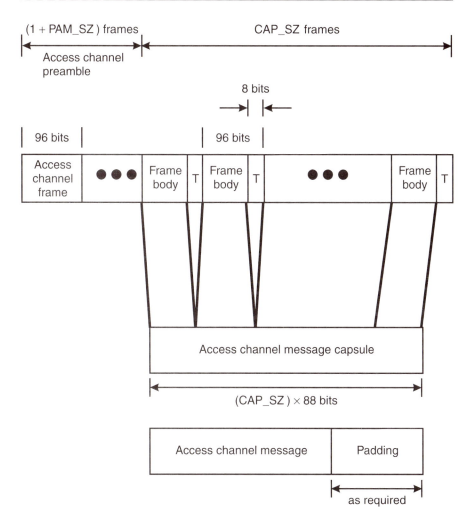

Figure 5.14 Access channel message structure.

traffic channel is similar to that of the access channel. The major difference is that the reverse traffic channel contains a data burst randomizer, as shown in Figure 5.15.

The orthogonally modulated data is fed into the data burst randomizer. The function of the data burst randomizer is to take advantage of the voice activity factor on the reverse link. Recall that the forward link uses a different scheme to take advantage of the voice activity factor—when the vocoder is operating at a lower rate, the forward link transmits the repeated symbols at a reduced energy per symbol and thereby reduces the forward-link power during

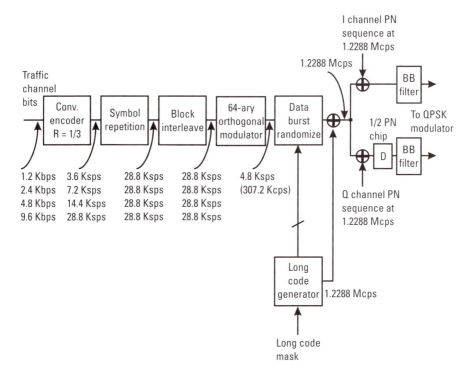

Figure 5.15 Reverse traffic channel for Rate Set 1. After[1].

any given period. As shown by Figure 5.9, at a reduced rate (i.e., 1 sps) and thus at a reduced energy, the integrator needs to integrate over a longer period of time (i.e., 1 sec) in order to accumulate enough power for proper symbol detection. In other words, at a reduced rate, the receiver takes longer to detect each symbol. This scheme is fine for a forward link where the speed requirement of forward-link power control is not stringent. After all, the mobile measures FER over a relatively longer period of time and reports this FER back to the base station. Then, the base station would act upon the information and adjust its forward power.

The scheme described for the forward link is inadequate for the reverse link. The reason is that the requirement of power-control speed is much more stringent on the reverse link. The base station measures the E_b/N_0 (a quantity that is quickly measured) on the reverse link, then the base station immediately makes a power-control decision and sends a PCB back to the mobile. The base station needs to quickly detect each symbol, even when the vocoder is operating at a lower rate, and the base station cannot afford to wait for the integrator to accumulate energy over a long period of time.

The approach taken to reduce reverse-link power during quieter periods of speech is to pseudorandomly mask out redundant symbols produced by symbol repetition. This is accomplished by the data burst randomizer. The data burst randomizer generates a masking pattern of 0s and 1s that randomly masks out redundant data. The masking pattern is partially determined by the vocoder rate. If the vocoder is operating at 9.6 Kbps, then no data is masked. If the vocoder is operating at 1.2 Kbps, then the symbols are repeated seven times, and the data burst randomizer masks out, on average, seven out of eight groups of symbols.

In actuality, each 20-ms traffic channel frame is divided into 16 power control groups, each 1.25 ms in length. The data burst randomizer pseudorandomly masks out individual power-control groups. When the vocoder operates at 9.6 Kbps, no PCG is masked out; when the vocoder operates at 4.8 Kbps, an average of 8 PCGs are masked out in a frame; when the vocoder operates at 2.4 Kbps, an average of 12 PCGs are masked out in a frame; and when the vocoder operates at 1.2 Kbps, an average of 14 PCGs are masked out in a frame. Figure 5.16 shows an example of this operation when the vocoder operates at 2.4 Kbps. In addition to depending on the vocoder rate, the masking pattern also depends on the long PN sequence used to spread the previous frame. For details of the masking algorithm, consult [1].

The reverse traffic channel structure is similar for Rate Set 2. The Rate Set 2 vocoder codes speech to data at a higher rate, and it delivers a better voice quality than that of Rate Set 1. The Rate Set 2 vocoder supports four variable rates: 14.4, 7.2, 3.6, and 1.8 Kbps. Figure 5.17 shows the reverse traffic channel for Rate Set 2. Note that in order to maintain the output of the block interleaver at 28.8 Ksps, the rate of the convolutional encoder is increased to $R = 1/2$.

5.4 Traffic Channel Formats

For both forward and reverse links, the traffic channel frames are 20 ms in duration. The frame at full rate contains 192 bits, yielding a rate of 9.6 Kbps; the frame at half rate contains 96 bits, yielding a rate of 4.8 Kbps; the frame at one-quarter rate contains 48 bits, yielding a rate of 2.4 Kbps; and the frame at one-eighth rate contains 24 bits, yielding a rate of 1.2 Kbps. The full-rate and half-rate frames contain frame quality indicator (CRC) bits, and all frames contain encoder tail bits (eight encoder tail bits per frame). Figure 5.18 shows the traffic channel frame structure [1].

We mentioned previously that voice, data, and messaging information may be transmitted over the traffic channel. This is true for both forward and

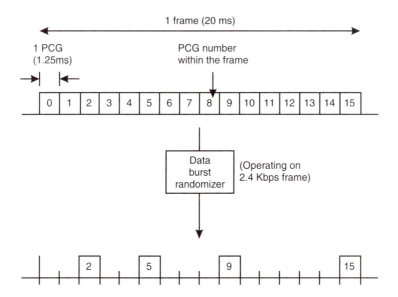

Figure 5.16 An example of data burst randomizer operation. After [1].

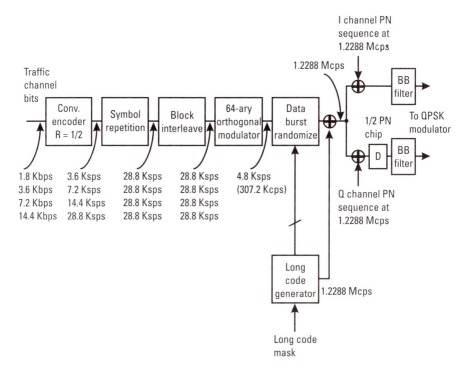

Figure 5.17 Reverse traffic channel for Rate Set 2.

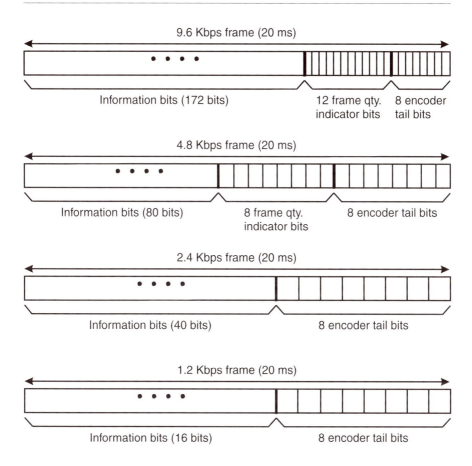

Figure 5.18 Traffic channel frame structure for both forward and reverse links. After [1].

reverse traffic channels. In fact, the system has the ability to multiplex *primary* and *signaling* or *secondary* data on the same traffic channel. The IS-95 CDMA system uses Multiplex Option 1 to transmit primary (i.e., voice) and secondary (i.e., data) traffic. This option is also used to transmit primary (i.e., voice) and signaling (i.e., messaging) traffic. Multiplex Option 1 uses the following techniques to simultaneously transmit primary and secondary or signaling traffic:

- *Blank and burst*: The entire traffic channel frame is used to send only secondary data. The entire traffic channel frame is also used to send only signaling data. The secondary or signaling data effectively blanks out the primary data.

- *Dim and burst*: The traffic channel frame is used to send both primary and secondary data. The traffic channel frame is also used to send both primary and signaling data.

It is important to note that these techniques are only used on those traffic channel frames that are full rate. In other words, secondary or signaling data is never multiplexed onto traffic channel frames that are half rate, one-quarter rate, or one-eighth rate. These lower rate frames contain primary traffic only.

Format bits in the beginning of a full-rate frame indicate what kind of traffic is contained in the frame. For example, if the first format bit of the frame is 0, then the frame is to be used for primary traffic only. If the first format bit of the frame is 1 and the following three format bits are 000, then 80 bits in the frame are to be used for primary traffic and 88 bits in the frame are to be used for signaling traffic. This is an example of dim and burst; note that in this case, the primary traffic is effectively sent at half rate (i.e., 80 bits per frame).

For a complete specification of Multiplex Option 1, consult Section 6.1.3.3.11 and Section 7.1.3.5.11 of [1].

5.4.1 Forward Link

When the mobile is communicating with the base station on a traffic channel, the base station may elect to send messages to the mobile while the traffic channel is still active. During traffic channel operation, the base station sends signaling messages to the mobile using the forward traffic channel [1]. The signaling information is sent over the traffic channel using Multiplex Option 1, described in the previous section. The base station may use one or more forward traffic channel frames to send a particular message.

The following are some important messages that the base station sends to the mobile on the forward traffic channel:

The *in-traffic system parameters message* contains important system configuration parameters such as the current handoff parameters for the mobile to use [1]:

- T_ADD—pilot detection threshold;
- T_DROP—pilot drop threshold;
- T_COMP—active set versus candidate set comparison threshold;
- T_TDROP—drop timer value;
- SRCH_WIN_A—search-window size for the active and candidate sets;

- SRCH_WIN_N—search-window size for the neighbor set;
- SRCH_WIN_R—search-window size for the remaining set;
- NGHBR_MAX_AGE—neighbor set maximum age.

The *handoff direction message* contains information that the mobile needs in order to commence communication with a new base station (or new base stations). In effect, the current base station transmits the following three fields for each member of the mobile's new active set [1]:

- PILOT_PN—pilot PN sequence offset index;
- PWR_COMB_IND—power-control symbol combining indicator;
- CODE_CHAN—code channel index indicating which Walsh function to use for the new traffic channel.

In addition, the base station updates the following handoff parameters in the *handoff direction message*:

- HDM_SEQ—handoff direction message sequence number;
- T_ADD—pilot detection threshold;
- T_DROP—pilot drop threshold;
- T_COMP—active set versus candidate set comparison threshold;
- T_TDROP—drop timer value;
- SRCH_WIN_A—search window size for the active and candidate sets

The *neighbor list update message* updates the mobile with a new list of neighbors. The message contains the NGHBR_PN parameter (neighbor pilot PN sequence offset index) for each pilot in the neighbor list. The *power control parameters message* updates the mobile parameters to use for forward power control:

- PWR_REP_THRESH—power-control reporting threshold;
- PWR_REP_FRAMES—power-control reporting frame count;
- PWR_THRESH_ENABLE—threshold report mode indicator;
- PWR_PERIOD_ENABLE—periodic report mode indicator;
- PWR_REP_DELAY—power report delay.

5.4.2 Reverse Link

When the mobile is communicating with the base station on a traffic channel, the mobile may send messages back to the base station while the traffic channel is still active. During traffic channel operation, the mobile sends signaling messages to the base station using the reverse traffic channel [1]. The signaling information is sent over the traffic channel using Multiplex Option 1, described in Section 5.4. The mobile may use one or more reverse traffic channel frames to send a particular message.

The following are some important messages that the mobile sends to the base station on the traffic channel.

The mobile sends the *pilot strength measurement message* informing the base station of the measured pilot strength for each member of its active and candidate sets [1]. The following parameters are reported for each member of the mobile's active and candidate sets:

- PILOT_PN_PHASE—pilot measured phase;
- PILOT_STRENGTH—pilot strength.

The *power measurement report message* is sent back to the base station to report error statistics measured on the forward link. This message is used by the base station for power control on the forward link. The following parameters are reported by the mobile [1]:

- ERRORS_DETECTED—number of frame errors detected;
- PWR_MEAS_FRAMES—number of forward traffic channel frames in the measurement period;
- PILOT_STRENGTH—pilot strength measured for each member of the active set.

The *handoff completion message* is sent back to the base station to report that the mobile has completed the handoff specified in a previous *handoff direction message*. The mobile reports the following parameters [1]:

- LAST_HDM_SEQ—*handoff direction message* sequence number that corresponds to the HDM_SEQ field from the *handoff direction message* that determined the current active set;
- PILOT_PN—pilot PN sequence offset for each member of the active set.

References

[1] TIA/EIA IS-95A, "Mobile Station-Base Station Compatibility Standard for Dual-Mode Wideband Spread Spectrum Cellular System," Telecommunications Industry Association.

[2] Garg, V. K., K. Smolik, and J. E. Wilkes, *Applications of CDMA in Wireless/Personal Communications*, Upper Saddle River, NJ: Prentice Hall, 1997.

Select Bibliography

Harte, L., *CDMA IS-95 for Cellular and PCS: Technology, Applications, and Resource Guide*, New York, NY: McGraw-Hill, 1997.

Labedz, G., et al., "Predicting Real-World Performance for Key Parameters in a CDMA Cellular System," *Proc. 46th Annual Vehicular Technology Conf.*, IEEE, 1996, pp. 1472–1476.

Lin, S., and D. J. Costello, Jr., *Error Control Coding: Fundamentals and Applications*, Englewood Cliffs, NJ: Prentice Hall, 1983.

Qualcomm, *The CDMA Network Engineering Handbook, Vol. 1: Concepts in CDMA*, 1993.

TIA PN-3570 (TSB-74), "Support for 14.4 Kbps Data Rate and PCS Interaction for Wideband Spread Spectrum Cellular Systems," Telecommunications Industry Association.

TIA IS-665, "W-CDMA Air Interface Compatibility Standard for 1.85-1.99 GHz PCS Applications."

TIA/EIA IS-96A, "Speech Service Option Standard for Wideband Spread Spectrum Digital Cellular System," Telecommunications Industry Association.

6

Call Processing

6.1 Call Processing States

Call processing refers to all the necessary functions that the system needs to carry out in order to set up, maintain, and tear down a call between a mobile and another party. Two types of connections are possible: a *mobile-to-land* call and a *mobile-to-mobile* call. In the case of mobile to land, the call is set up between a mobile and a landline phone, in which case the call is routed through the *public switched telephone network* (PSTN). The IS-95 CDMA system adopts a state description of call processing. Since the mobile is the common element in the two types of connections (i.e., mobile to land and mobile to mobile), the IS-95 standard specifies the call states from the perspective of the CDMA mobile station.

It is important to note that the standard does not specify call states for the base station. Obviously, whatever functions that the base station performs must work with the specified mobile call states; the infrastructure vendors are free to implement their own base station functions to satisfy call processing requirements [1].

During normal operation, the mobile can occupy any one of the following states [2]:

- Mobile station initialization state;
- Mobile station idle state;
- System access state;
- Mobile station control on the traffic channel state.

Figure 6.1 graphically depicts these states and the associated transitions. After power-up, the mobile first enters the mobile station initialization state (or initialization state for short), where the mobile selects and acquires a system. Upon exiting the initialization state, the mobile has fully acquired the system and its timing. Then the mobile enters the mobile station idle state (or idle state for short), where the mobile monitors messages on the paging channel [2].

Any one of the following three events will cause the mobile to transition from the idle state to the system access state (or access state for short): 1) the mobile receives a paging channel message requiring an acknowledgment or response, 2) the mobile originates a call, or 3) the mobile performs a registration. In the access state, the mobile sends messages to the base station on the access channel. When the mobile is directed to a traffic channel, it enters the mobile station control on the traffic channel state (or traffic channel state for short), where the mobile communicates with the base station using the forward

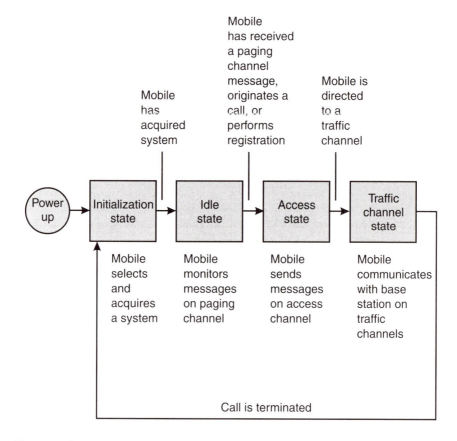

Figure 6.1 Call processing states of the mobile station. After [2].

and reverse traffic channels. When the call is terminated, the mobile returns to the initialization state [2].

6.2 Initialization State

After power-up, the mobile enters the initialization state. This state contains four *substates*, which the mobile sequentially goes through:

1. System determination substate;
2. Pilot channel acquisition substate;
3. Sync channel acquisition substate;
4. Timing change substate.

6.2.1 System Determination Substate

The system determination substate is the first substate that the mobile enters within the initialization state. In this substate, the mobile selects which system to use (i.e., system A or system B in the cellular band). Because all CDMA cellular phones have dual-mode capability, the mobile also decides whether it would be in CDMA mode or analog mode. If CDMA mode is selected, the mobile proceeds to select which CDMA carrier to use. Then the mobile enters the next substate.

6.2.2 Pilot Channel Acquisition Substate

In the pilot channel acquisition substate, the mobile demodulates and acquires the pilot channel of the selected CDMA system. See Sections 5.2.1 and 5.2.2 for details on pilot channel synchronization. The mobile has to acquire the pilot within a certain time limit; if the mobile acquires the pilot within the time limit, then it enters the sync channel acquisition substate. If the mobile does not acquire the pilot within the time limit, then it returns to the system determination substate.

6.2.3 Sync Channel Acquisition Substate

In the sync channel acquisition substate, the mobile proceeds to acquire the sync channel of the CDMA system. In effect, the mobile obtains system configuration and timing information by acquiring the sync channel and reading the *sync channel message*. See Section 5.2.2 for details on sync channel synchronization. Note that the mobile also needs to acquire the sync channel and read the message within a certain time limit. If the mobile does not receive the *sync channel message* within the time limit, then it returns to the system determination substate. If the mobile receives the *sync channel message* within the time allowed, and if the mobile's own protocol revision level (MOB_P_REV) is greater than or equal to the minimum protocol revision level supported by the base station (MIN_P_REV), then the mobile proceeds to enter the next substate.

6.2.4 Timing Change Substate

In the timing change substate, the mobile synchronizes its timing to that of the CDMA system time and its long-code phase to that of the CDMA system. At this point, the mobile has already demodulated the *sync channel message* and possesses all the parameters from the message; three parameters in the message (i.e., PILOT_PN, LC_STATE, and SYS_TIME) are used to synchronize the mobile's long-code phase and system timing to those of the CDMA system.

After the mobile has fully acquired the CDMA system, it then enters the mobile station idle state.

6.3   Idle State

6.3.1   Paging Channel Monitoring

In the idle state, the mobile monitors the paging channel on the forward link. In order to receive messages and receive an incoming call, the mobile needs to monitor the paging channel for messages. The paging channel transmission is divided into slots that are 80 ms in length (see Section 5.2.3). There are two ways that the mobile can monitor the paging channel: monitoring in *nonslotted mode* or in *slotted mode*.

In nonslotted mode, the mobile monitors the paging channel at all times [2].

In slotted mode, the mobile monitors the paging channel only during assigned paging channel slots. Because the mobile doesn't have to monitor all the slots all the time, the mobile operating in the slotted mode can conserve battery power.

6.3.2 Idle Handoff

When the mobile is in the idle state and it has moved from the coverage area of one base station to the coverage area of another base station, an *idle handoff* occurs. If the mobile detects that the pilot strength from another base station is sufficiently stronger than that of the current base station, then the mobile proceeds to perform an idle handoff. In doing so, the mobile maintains three sets of base station (or sector) pilot PN offsets:

- *Active set*: This set contains the pilot offset of the sector whose paging channel is currently monitored by the mobile.
- *Neighbor set*: This set contains the pilot offsets of those sectors that are likely candidates for idle handoff. The *neighbor list message* specifies the pilots in the neighbor set.
- *Remaining set*: This set contains all possible pilot offsets in the system, excluding the pilots in the active set and the neighbor set [2].

The mobile also uses three search windows—SRCH_WIN_A, SRCH_WIN_N, and SRCH_WIN_R—to search for those pilots contained in the respective sets. If the mobile determines that one of the neighbor set or remaining set pilot signals is sufficiently stronger than the pilot of the active set, the mobile performs an idle handoff. The idle handoff is nothing more than beginning to monitor the paging channel of a new sector whose pilot strength is sufficiently stronger than that of the current sector [2].

It is important to recognize that the mobile monitors the paging channel of only one base station (i.e., the active set contains only one pilot). Therefore, soft handoff is not applicable in the idle state.

6.3.3 Paging Channel Messages

There is a total of six overhead messages that are sent to the mobile on the paging channel:

- *System parameters message;*
- *Neighbor list message;*
- *CDMA channel list message;*
- *Extended system parameters message;*
- *Global service redirection message;*
- *Access parameter message.*

The first five messages are referred to as the *configuration* messages. The mobile is constantly receiving these various messages on the paging channel, and within these messages there are many fields that need to be updated and loaded into the mobile's memory. So, how does the mobile keep track of which messages are current and which messages are not?

It turns out that the base station assigns a configuration message sequence number (CONFIG_MSG_SEQ) to a current set of configuration messages. When the contents (fields) of one or more of the configuration messages changes, the configuration message sequence number is incremented [2].

For each of the configuration messages, the mobile keeps or stores locally a message sequence number pertaining to that particular message:

- SYS_PAR_MSG_SEQ$_S$, or stored system parameters message sequence number;

- NGHBR_LIST_MSG_SEQ$_S$, or stored neighbor list message sequence number;

- CHAN_LIST_MSG_SEQ$_S$, or stored CDMA channel list message sequence number;

- EXT_SYS_PAR_MSG_SEQ$_S$, or stored extended system parameters message sequence number;

- GLOB_SERV_REDIR_MSG_SEQ$_S$, or stored global service redirection message sequence number.

For each received configuration message, the mobile stores the configuration message sequence number (contained in the configuration message) in the respective message sequence number. For example, if the CONFIG_MSG_SEQ field of a received neighbor list message is 11, then the mobile should update its locally stored NGHBR_LIST_MSG_SEQ$_S$ to 11.

The mobile considers its stored configuration parameters to be current only if all the stored message sequence numbers; that is, SYS_PAR_MSG_SEQ$_S$, NGHBR_LIST_MSG_SEQ$_S$, CHAN_LIST_MSG_SEQ$_S$, EXT_SYS_PAR_MSG_SEQ$_S$, and GLOB_SERV_REDIR_MSG_SEQ$_S$, are equal to the stored CONFIG_MSG_SEQ$_S$. The stored CONFIG_MSG_SEQ$_S$ is simply the most recently received configuration message sequence number of all the configuration messages.

The access parameters message is sequence-numbered by its ACC_MSG_SEQ field, and to keep track of the most current message, the mobile stores the most recently received access parameters message sequence number ACC_MSG_SEQ$_S$.

Refer to Section 5.2.3 for the contents of those messages that relate to RF system engineering.

6.4 Access State

In the access state, the mobile transmits messages to the base station using the access channel. In addition, the mobile also receives messages from the base station on the paging channel. There are six substates that the mobile can occupy within the access state [2]:

- Update overhead information substate;
- Page response substate;
- Mobile station origination attempt substate;
- Registration access substate;
- Mobile station order/message response substate;
- Mobile station message transmission substate.

6.4.1 Update Overhead Information Substate

After the mobile receives the current configuration messages on the paging channel, the mobile compares the sequence numbers to determine whether or not all the configuration messages are up to date [2]. Update procedures (described in Section 6.3.3) are used to update all the locally stored sequence numbers. The mobile also checks whether or not it has the latest access parameters by checking its locally stored access parameters message sequence number $ACC\_MSG\_SEQ_S$.

In addition to receiving configuration messages and *access parameter messages*, the mobile can also receive the following page messages:

- *Page message;*
- *Slotted page message;*
- *General page message.*

Whenever the mobile receives a page message, the mobile station searches each message to determine whether the message contains the mobile's *international mobile station identity* (IMSI). If the message contains the mobile's IMSI, then the mobile transitions to the page response substate and transmits a *page response message* on the access channel.

6.4.2   Page Response Substate

In this substate, the mobile sends a *page response message* in response to the page messages sent by the base station [2]. The mobile sends the page response message using the access procedures described in Section 6.4.7.

After receiving the page response message, the base station may send the mobile a *channel assignment message* on the paging channel in order to start setting up the call. The channel assignment message contains parameters such as CDMA_FREQ (frequency assignment) and CODE_CHAN (code channel), and these parameters are used by the mobile to tune to the assigned RF frequency and CDMA code channel in order to start receiving the forward traffic channel.

6.4.3   Mobile Station Origination Attempt Substate

In this substate, the mobile sends an *origination message* to the base station in order to originate a call [2]. The mobile sends the origination message using the access procedures described in Section 6.4.7.

After receiving the *origination message*, the base station may send the mobile a *channel assignment message* on the paging channel in order to start setting up the call. The channel assignment message contains parameters such as CDMA_FREQ (frequency assignment) and CODE_CHAN (code channel), and these parameters are used by the mobile to tune to the assigned RF frequency and CDMA code channel in order to start receiving the forward traffic channel.

6.4.4   Registration Access Substate

In this substate, the mobile sends a *registration message* to the base station [2]. The mobile sends the *registration message* using the access procedures described in Section 6.4.7.

Registration is the process where the mobile informs the base station about the mobile's identification, status, location, and other pertinent information. For example, the mobile may inform the base station of the mobile's location so that the system can efficiently page the mobile when there is an incoming call [2].

6.4.5   Mobile Station Order/Message Response Substate

In this substate, the mobile sends a response to any other message sent by the base station [2]. For example, the mobile may send an *authentication challenge response message* in response to an *authentication challenge message* sent by the base station. The mobile sends the pertinent message using the access procedures described in Section 6.4.7.

6.4.6 Mobile Station Message Transmission Substate

A mobile phone does not have to support this particular substate. In other words, the mobile's support of this substate is optional. In this substate, the mobile sends a *data burst message* to the base station. The mobile sends the data burst message using the access procedures described in Section 6.4.7.

6.4.7 Access Procedures

When the mobile sends messages to the base station on the access channel, there is a fundamental problem of accommodating more than one mobile. For example, there are typically many mobiles that wish to access one base station. To some extent, this *congestion* problem can be solved by separating mobiles by assigning them different access channels to use. The base station does this by specifying the parameter ACC_CHAN in the *access parameters message*; the mobile randomly picks an access channel between 0 and ACC_CHAN to transmit.

However, if more than one mobile uses the same access channel, their access channel transmissions might collide in time. This *collision* problem can also be solved by randomizing the transmission time of the different mobiles such that the probability of collision is reduced.

There are two types of messages sent on the access channel: a response message and a request message. A response message is sent in response to a base station message (e.g., an *authentication challenge response message* is an example of a response message). On the other hand, a request message is sent autonomously by the mobile (e.g., an *origination message* is an example of a request message). Different access procedures are used to send these two types of messages [2].

Figure 6.2 depicts an *access attempt* by the mobile that is sending a transmission on the access channel. Several observations can be made regarding the access attempt:

1. An access attempt consists of several *access probe sequences*. For sending response messages, the maximum number of sequences allowed in the access attempt is given by the MAX_RSP_SEQ parameter (i.e., $N \leq$ MAX_RSP_SEQ. MAX_RSP_SEQ is a parameter that can be

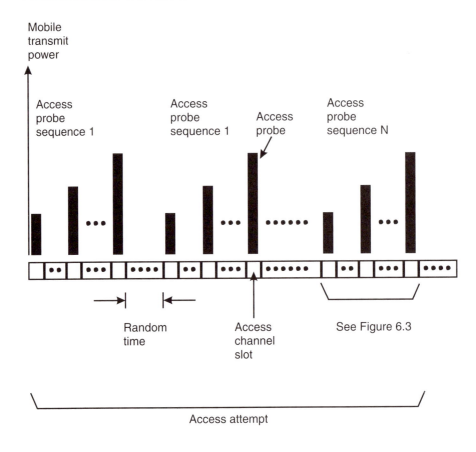

Figure 6.2 Access attempt.

changed by the system operator). Note that MAX_RSP_SEQ itself has an upper limit of 15.

2. For sending request messages, the maximum number of sequences allowed in the access attempt is given by the MAX_REQ_SEQ parameter (i.e., $N \leq$ MAX_REQ_SEQ. MAX_REQ_SEQ is a parameter that can be changed by the system operator). Note that MAX_REQ_SEQ itself has an upper limit of 15.

3. A new access channel is randomly generated between 0 and ACC_CHAN before every access probe sequence. So, in effect, each access probe sequence can be transmitted on a different access channel.

4. There is a random time between access probe sequences. For sending response messages, the random time is a random number between 0 and (1 + BKOFF) in the unit of access channel slots. For example, if BKOFF is 3, then a random number is generated between 0 and 4. Say the random number generated is 3; then, the random time between the two access probe sequences is 3 slots. Note that this random time would be different after each access probe sequence (i.e., a random number is generated after each access probe sequence). This random number is also known as the sequence backoff.

5. The random time for request messages is generated differently. In addition to the random number between 0 and (1 + BKOFF), there is an additional *persistence delay* embedded in the random time. In other words, the random time is the sum of the random number and the persistence delay, all in the unit of access channel slots. The actual amount of persistence delay depends on the outcome of a *persistence test* performed by the mobile; the test uses the reason for access and class of service as inputs, and has a pseudorandom outcome. For example, if the priority of access is high, then it is more likely that the persistence test result is a pass. Therefore, after the transmission of the current access probe sequence, the mobile would first delay transmission by a random number between 0 and (1 + BKOFF), then the mobile would perform the persistence test. If the test passes, then the next access probe sequence would begin immediately; if the test fails, then the next access probe sequence would be delayed by one slot. Then the mobile would perform the persistence test again. The persistence delay continues slot by slot until the persistence test passes.

Persistence delay is used to further randomize the transmission time of request messages (i.e., those messages that are transmitted autonomously by the mobile). This is needed because the base station cannot control the rate of arrival of request messages. The persistence delay is not needed in transmitting response messages because the base station can indirectly control the arrival rate of response messages; the base station does so by controlling the rate it transmits messages requiring responses [2].

Figure 6.3 depicts an access probe sequence in detail. Some observations can also be made with the access probe sequence:

1. An access probe sequence consists of several *access probes*. An access probe is basically a transmission contained in one access channel slot, and the transmission is sent at some power level. The maximum

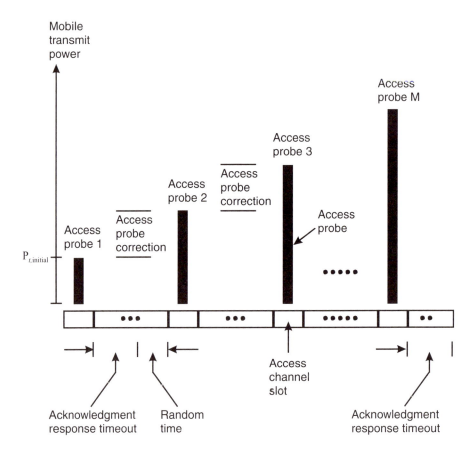

Figure 6.3 Access probe sequence.

number of probes allowed in an access probe sequence is given by the NUM_STEP parameter (i.e., $M \le (\text{NUM\_STEP} + 1)$). NUM_STEP is a parameter that can be changed by the system operator. Note that NUM_STEP itself has an upper limit of 15.

2. The initial power of the first access probe is controlled by the open-loop power control (see Section 4.3.2). The initial power is given by (4.15) and reproduced here:

$$p_{t,\text{initial}} = -p_r - 73 + \text{NOM\_PWR} + \text{INIT\_PWR}$$

3. The powers of successive access probes are incremented by an amount known as the access probe correction (see Section 4.3.2). The access

probe correction is equal to that specified by the parameter PWR_STEP.

4. After the transmission of an access probe, the mobile will wait for a period of time known as the acknowledgment response timeout. If an acknowledgment from the base station is received during this time, then the access attempt is successful and the mobile ends the access attempt. The acknowledgment response timeout is a deterministic number generated by using the parameter ACC_TMO. Specifically, the acknowledgment response timeout, in milliseconds, is equal to $80 \times (2 + ACC\_TMO)$.

5. If no acknowledgment is received, the mobile waits an additional random time before transmitting the next access probe. This random time is a random number generated between 0 and $(1 + PROBE\_BKOFF)$ in the unit of access channel slots.

6.5 Traffic Channel State

The mobile may enter the traffic channel state from two substates within the access state: the page response substate or the mobile station origination attempt substate. In other words, after the mobile has successfully responded to a base station page or after the mobile has successfully originated, the mobile may enter the traffic channel state. In the traffic channel state, the mobile communicates with the base station using the forward and reverse traffic channels [2]. This state consists of five substates:

- Traffic channel initialization substate;
- Waiting for order substate;
- Waiting for mobile station answer substate;
- Conversation substate;
- Release substate.

6.5.1 Traffic Channel Initialization Substate

In the traffic channel initialization substate, the mobile checks to see if it can indeed receive information on the forward traffic channel. In doing so, the mobile verifies that it can receive two (i.e., N_{5m} constant defined in IS-95A) consecutive good frames on the forward traffic channel within 200 ms (i.e., T_{50m} constant defined in IS-95A). If it can, then the mobile begins to transmit on the reverse traffic channel (i.e., the mobile starts to transmit the traffic channel preamble on the reverse traffic channel).

If the mobile receives a base station acknowledgment order from the base station within two seconds (i.e., T_{51m} constant defined in IS-95A) after entering this substate, then the following will occur:

- If the call is mobile terminated, then the mobile would enter the wait for order substate.

- If the call is mobile originated, then the mobile would enter the conversation substate.

Of course, things may go wrong while the mobile is in this substate. If the mobile does not receive two consecutive good frames within 200 ms, or if the mobile does not receive a base station acknowledgment order within two seconds after entering this substate, then the mobile would return to the system determination substate of the initialization state.

6.5.2 Waiting for Order Substate

If the call is mobile terminated, the mobile enters this substate from the traffic channel initialization substate. In this substate, the mobile waits for an *alert with information message* (sent on the forward traffic channel) from the base station. This message basically conveys an alert, or ring, to the mobile. If the mobile receives the alert message, then the mobile would enter the waiting for mobile station answer substate.

If the mobile does not receive the alert message within five seconds after entering this substate, then the mobile would return to the system determination substate of the initialization state.

6.5.3 Waiting for Mobile Station Answer Substate

Normally, the mobile enters this substate from the waiting for order substate. In this substate, the mobile waits for the user to answer the call. Note that the mobile can only enter this substate if the call is mobile terminated. If the user answers the call, then the mobile stops ringing and sends a connect order back to the base station on the reverse traffic channel; then, the mobile enters the conversation substate.

6.5.4 Conversation Substate

If the call is mobile originated, then the mobile enters the conversation substate from the traffic channel initialization substate. If the call is mobile terminated, then the mobile enters this substate from the waiting for mobile station answer substate. In this substate, the mobile and the base station exchange primary traffic bits on forward and reverse traffic channels.

6.5.5 Release Substate

In the release substate, the mobile releases or disconnects the call. If the mobile initiates the release, then the mobile first sends a release order to the base station on the reverse traffic channel. If the base station initiates the release, then the base station sends a release order to the mobile on the forward traffic channel.

References

[1]   Garg, V. K., K. Smolik, and J. E. Wilkes, *Applications of CDMA in Wireless/Personal Communications*, Upper Saddle River, NJ: Prentice Hall, 1997.

[2]   TIA/EIA IS-95A, "Mobile Station-Base Station Compatibility Standard for Dual-Mode Wideband Spread Spectrum Cellular System," Telecommunications Industry Association.

Select Bibliography

Harte, L., *CDMA IS-95 for Cellular and PCS: Technology, Applications and Resource Guide*, New York, NY: McGraw-Hill, 1997.

TIA IS-665, "W-CDMA Air Interface Compatibility Standard for 1.85-1.99 GHz PCS Applications," Telecommunications Industry Association.

7

CDMA Design Engineering

7.1 Introduction

Before implementing a cell, a cluster of cells, or a system, RF engineers need to know whether or not the CDMA design of the cell, cluster, or system will support the basic radio links. In other words, are all the radio parameters adequate to maintain a high-quality radio link between the base station and the mobile?

We presented in Chapter 2 the link equation that characterized the quality of a radio link. Specifically, the carrier-to-noise ratio was treated as the figure of merit for the link. In this section, we adapt the link equation for use in the analysis of the forward and reverse CDMA links. There are three important parameters to consider in CDMA design: the E_c/I_0 of the pilot channel, the E_b/N_0 of the forward traffic channel, and the E_b/N_0 of the reverse traffic channel. Note that throughout Chapter 7, we refer to the mobile of interest as the *probe mobile*, and we examine mathematical expressions of forward link E_c/I_0, forward link E_b/N_0, and reverse link E_b/N_0 from the perspective of this probe mobile.

7.2 Forward Link Analysis

7.2.1 Pilot Channel

The E_c/I_0 is the energy per chip per interference density measured on the pilot channel; it is effectively the signal strength of the pilot channel. The mobile continuously measures the E_c/I_0 and compares it against the different thresholds, such as the pilot detection threshold T_ADD and the pilot drop threshold T_DROP. The results of these comparisons are reported back to the base station so that the base station can make a determination of whether or not the mobile should be handed off from one base station to the next. Thus, the E_c/I_0 plays a prominent role in determining whether or not a mobile is within the coverage area of a base station. Furthermore, the pilot signal is transmitted by a base station at a relatively higher power than those of other forward-link logical channels. A call cannot be set up without the mobile's successful reception of the pilot channel because, along with other functions, the pilot channel serves as a coherent carrier phase reference for demodulation of other logical channels on the forward link. Therefore, the E_c/I_0 effectively determines the forward coverage area of a cell or sector, and one has to ensure that the forward link E_c/I_0 strength is sufficient. In developing an expression for the E_c/I_0, we shall consider four different cases of increasing complexity:

- Single cell and single mobile;
- Many cells and single mobile;
- Single cell and many mobiles;
- Many cells and many mobiles.

7.2.1.1 Single Cell and Single Mobile

In this situation, the probe mobile is receiving the pilot signal from only one base station (see Figure 7.1), but in addition to thermal noise, the total overhead power from this base station is also interfering with the pilot signal. Recall that all logical channels are transmitted in the same RF band. For a single cell serving a single mobile, the E_c/I_0 measured by the probe mobile is given by

$$\frac{E_c}{I_0} = \frac{\alpha_0 P_0(\theta_0) L_0(\theta_0, d_0) G}{I_h + I_n + N} \tag{7.1}$$

where

- $P_0(\theta_0)$ = home base station (sector 0) overhead ERP including pilot, paging, and sync powers in the direction θ_0 to the probe mobile. Note that in general—because ERP depends on the antenna pattern, which is a function of direction θ_0—ERP itself is also a function of direction θ_0.

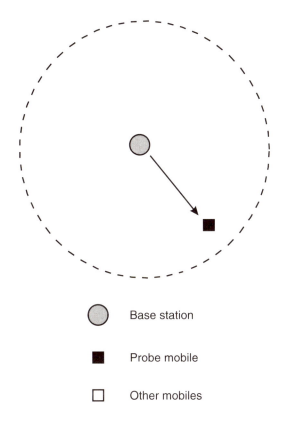

Figure 7.1 A single cell serving a single mobile—forward-link characterization.

- α_0 = fraction of home base station overhead ERP allocated to pilot power.

- $L_0(\theta_0,d_0)$ = path loss from home base station in the direction θ_0 to the probe mobile a distance d_0 away.

- G = receive antenna gain of probe mobile.

- I_h = power received at the probe mobile from overhead power emitted by home base station.

- I_n = power received at the probe mobile from other interference of non-CDMA origins. This term is included to accommodate all other possible interference sources that could be jamming the system in the CDMA band.

- N = thermal noise power.

Equation (7.1) assumes that the single forward traffic channel assigned to the probe mobile is either not yet active or negligible, and thus we do not include the interference contribution from this single traffic channel in the denominator. We also include the interference term I_n to take care of interference from other miscellaneous sources.

For the case of a single base station and a single probe mobile, the overhead (interference) power received (i.e., I_h) at the probe mobile is deterministic and given by

$$I_h = P_0(\theta_0)L_0(\theta_0, d_0)G \tag{7.2}$$

where all of the overhead power $P_0(\theta_0)$ from the base station contributes to the interference term I_h in the denominator.

7.2.1.2 Many Cells and Single Mobile

For the case of many base stations and a single probe mobile, the situation is similar to that described in the previous section. However, in addition to receiving the deterministic overhead powers from the home base station, the probe mobile is also intercepting the deterministic overhead powers from other surrounding base stations (see Figure 7.2). These overhead powers from other base stations represent an additional term of interference in the denominator. Therefore, (7.1) can be modified to

$$\frac{E_c}{I_0} = \frac{\alpha_0 P_0(\theta_0)L_0(\theta_0, d_0)G}{I_h + I_n + I_o + N} \tag{7.3}$$

where I_o is the sum of overhead powers from other base stations. I_o is deterministic and is given by

$$I_o = G\sum_{k=1}^{K} P_k(\theta_k)L_k(\theta_k, d_k) \tag{7.4}$$

Equation (7.4) is simply a summation of all overhead powers received at the probe mobile (excluding the overhead power from home base station $k = 0$). Note that there is a total of K base stations (or sectors) in the system.

7.2.1.3 Single Cell and Many Mobiles

In this situation, a single base station is serving many mobiles within its coverage area (including the probe mobile) and there are no other base stations

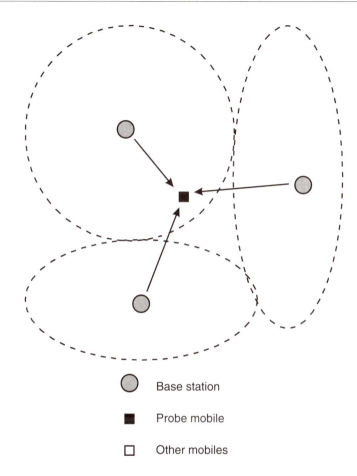

○ Base station

■ Probe mobile

□ Other mobiles

Figure 7.2 A single cell serving a single mobile—forward-link characterization. In this case, the mobile is also receiving powers from surrounding cells.

around (see Figure 7.3). Therefore, all mobiles are receiving their respective forward traffic channel powers from the base station, but these traffic channel powers represent interference to the probe mobile. Equation (7.1) can be modified to include this additional interference term; that is,

$$\frac{E_c}{I_0} = \frac{\alpha_0 P_0(\theta_0) L_0(\theta_0, d_0) G}{I_h + I_n + I_m + N} \tag{7.5}$$

where I_m is the total traffic channel power (from the home base station) received at the probe mobile (which is measuring the E_c/I_0). In other words, I_m is given by

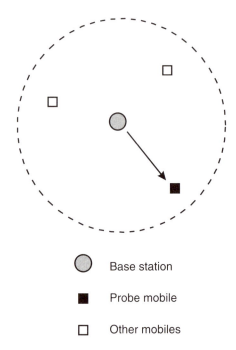

Figure 7.3 A single cell serving many mobiles—forward-link characterization.

$$I_m = GL_0(\theta_0, d_0)\sum_{j=1}^{J}T_j(\theta_0)$$ (7.6)

where $T_j(\theta_0)$ is the forward traffic channel ERP intended for mobile j but radiated in the direction θ_0 toward the probe mobile (also designated as mobile 0). $T_j(\theta_0)$ can be interpreted as the traffic channel ERP intended for mobile j but intercepted by mobile 0. Note that J is the total number of mobiles being served by the base station in question.

In this case, I_m is not deterministic. Because of forward power control, the base station is constantly adjusting its transmit power on a particular traffic channel. Since the power of each forward traffic channel (i.e., T_j) is random, the sum of powers for all traffic channels is also random. In practice, one may either assume a nominal value for I_m or perform a Monte Carlo simulation to assess the simulated value of I_m.

7.2.1.4 Many Cells and Many Mobiles

In this case, there are many cells serving many mobiles in their respective coverage areas (see Figure 7.4). The E_c/I_0 of the probe mobile is given by

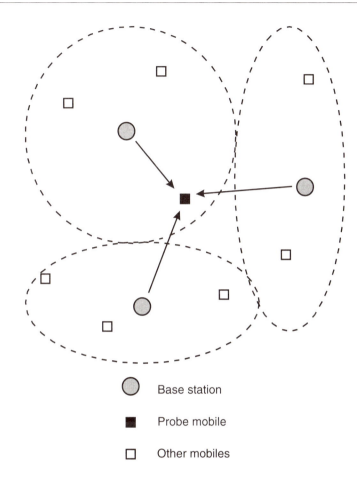

Figure 7.4 Many cells serving many mobiles—forward-link characterization.

$$\frac{E_c}{I_0} = \frac{\alpha_0 P_0(\theta_0) L_0(\theta_0, d_0) G}{I_b + I_n + I_o + I_m + I_t + N} \tag{7.7}$$

In addition to containing interference terms I_b, I_n, I_o, I_m, and N described in previous sections, (7.7) has an additional interference term I_t in the denominator. I_t is the total traffic channel power (received at the probe mobile) from all other base stations. In other words, I_t is given by

$$I_t = G \sum_{k=1}^{K} X_k(\theta_k) L_k(\theta_k, d_k) \tag{7.8}$$

where $X_k(\theta_k)$ is the total traffic channel ERP from base station k. Equation (7.8) is simply a summation of all the traffic channel powers (received at the probe mobile) from all other base stations (excluding home base station $k = 0$). Note that K designates the total number of cells or sectors in the system. Since $X_k(\theta_k)$ is the total traffic channel ERP from base station k, it is given by the following summation:

$$X_k(\theta_k) = \sum_{j=1}^{J_k} T_{k,j}(\theta_k)$$

(7.9)

Equation (7.9) states that for each base station k, we sum the forward traffic channel ERPs for all mobiles belonging to base station k. For base station k, $T_{k,j}(\theta_k)$ is the traffic channel ERP intended for mobile j but radiated in the direction θ_k toward the probe mobile. In other words, $T_{k,j}(\theta_k)$ can be interpreted as the traffic channel ERP transmitted by base station k intended for mobile j of that base station but intercepted by the probe mobile (which is measuring the E_c/I_0). Equation (7.9) sums over all such mobiles served by base station k, and J_k is the total number of mobiles served by base station k. Again, due to forward power control, $T_{k,j}(\theta_k)$ is random, and thus $X_k(\theta_k)$ is random, and consequently I_t is also random.

7.2.2 Traffic Channel

On the link level, we are concerned with the E_b/N_0 of the forward-link traffic channel. As shown in Chapter 3, the link E_b/N_0 translates directly into BER, which has implications on forward-link voice quality. Making sure that the link supports an adequate E_b/N_0 ensures the quality of the link. In order to illustrate the concept of link analysis without introducing complicating factors, we will not consider the effects of diversity gain in soft/softer handoff situations. We again consider four different cases:

- Single cell and single mobile;
- Many cells and single mobile;
- Single cell and many mobiles;
- Many cells and many mobiles.

7.2.2.1 Single Cell and Single Mobile

In this situation, the probe mobile is receiving power on the forward traffic channel from only one base station. For a single cell serving a single mobile, the E_b/N_0 received by that probe mobile is given by

$$\frac{E_b}{N_0} = \frac{T_0(\theta_0)L_0(\theta_0,d_0)G}{I_b + I_n + N}\left(\frac{W}{R}\right) \qquad (7.10)$$

where

- $T_0(\theta_0)$ home base station (sector 0) traffic channel ERP in the direction θ_0 to the probe mobile 0. Note that in general—because traffic channel ERP depends on the antenna pattern, which is a function of direction θ_0—ERP itself is also a function of direction θ_0.
- $L_0(\theta_0,d_0)$ = path loss from home base station in the direction θ_0 to the probe mobile a distance d_0 away.
- G = receive antenna gain of probe mobile.
- I_n = power received at the mobile station from other interference of non-CDMA origins. This term is included to accommodate all other possible interference sources that could be jamming the system in the CDMA band.
- N = thermal noise power.
- (W/R) = processing gain.

I_b is the interference power received at the probe mobile from the overhead power emitted by the home base station. It is given by

$$I_b = (1-\varepsilon)P_0(\theta_0)L_0(\theta_0,d_0)G \qquad (7.11)$$

where ε is the orthogonality factor. Note that due to forward power control, $T_0(\theta_0)$ is not fixed. Instead, $T_0(\theta_0)$ is constantly adjusted by the base station to maintain an acceptable link quality.

7.2.2.2 Many Cells and Single Mobile

This situation is analogous to that described in Section 7.2.1.2, where the probe mobile is intercepting overhead powers from surrounding base stations in addition to receiving overhead and traffic powers from the home base station; an additional interference term I_o in the denominator is necessary to describe these overhead powers from surrounding base stations. Therefore, (7.10) can be modified as

$$\frac{E_b}{N_0} = \frac{T_0(\theta_0)L_0(\theta_0,d_0)G}{I_b + I_n + I_o + N}\left(\frac{W}{R}\right) \qquad (7.12)$$

where I_o is deterministic and given by (7.4).

7.2.2.3 Single Cell and Many Mobiles

This case is similar to that described in Section 7.2.1.3 where there is only one base station in the system, and this single base station is serving many other mobiles in addition to serving the probe mobile. From the probe mobile's perspective, it is intercepting forward traffic channel transmissions intended for these other mobiles. Therefore, an interference term $I_{m'}$ is added in the denominator to account for the additional interference. Equation (7.10) is modified to include this additional term:

$$\frac{E_b}{N_0} = \frac{T_0(\theta_0)L_0(\theta_0, d_0)G}{I_b + I_n + I_{m'} + N}\left(\frac{W}{R}\right) \tag{7.13}$$

where $I_{m'}$ is the effective total traffic channel interference (from the home base station) intercepted by the probe mobile. $I_{m'}$ is given as

$$I_{m'} = (1 - \varepsilon)GL_0(\theta_0, d_0)\sum_{j=1}^{J}T_j(\theta_0) \tag{7.14}$$

where ε is the orthogonality factor, and $T_j(\theta_0)$ is the forward traffic channel ERP intended for mobile j but radiated in the direction θ_0 toward the probe mobile (which is measuring the E_b/N_0). The value $T_j(\theta_0)$ can also be interpreted as the traffic channel ERP intended for mobile j but intercepted by mobile 0. Note that $I_{m'}$ is not deterministic.

7.2.2.4 Many Cells and Many Mobiles

This case is analogous to that described in Section 7.2.1.4 where there are many base stations serving many mobiles. The E_b/N_0 at the probe mobile is given by

$$\frac{E_b}{N_0} = \frac{T_0(\theta_0)L_0(\theta_0, d_0)G}{I_b + I_n + I_o + I_{m'} + I_t + N}\left(\frac{W}{R}\right) \tag{7.15}$$

where the interference terms $I_b, I_n, I_o, I_{m'}, I_t$, and N are described in previous sections.

7.3 Reverse Link

Since there is no pilot channel on the reverse link, we are only interested in the E_b/N_0 of the reverse traffic channel. The link E_b/N_0 translates directly into

BER, which has implications on reverse-link voice quality. Making sure that the link supports an adequate E_b/N_0 ensures the quality of the link. In order to illustrate the concept of link analysis without introducing complicating factors, we will not consider the effects of diversity gain in soft/softer handoff situations. In developing an expression for the reverse E_b/N_0, we consider three different cases of increasing complexity:

- Single cell and single mobile;
- Single cell and many mobiles;
- Many cells and many mobiles.

7.3.1 Traffic Channel

7.3.1.1 Single Cell and Single Mobile

For a single cell serving a single mobile, the expression for the reverse link E_b/N_0 is similar to that of the forward link E_b/N_0 given in (7.10). The reverse link E_b/N_0 is given by

$$\frac{E_b}{N_0} = \frac{T'L'_0(\theta_0, d_0)G_0(\theta_0)}{I'_n + N}\left(\frac{W}{R}\right) \tag{7.16}$$

- T' = reverse traffic channel ERP of the probe mobile; the transmit pattern is assumed to be omnidirectional.
- $L'_0(\theta_0, d_0)$ = reverse path loss from the probe mobile in the direction θ_0 to the home base station a distance d_0 away.
- $G_0(\theta_0)$ = receive antenna gain of home base station in the direction θ_0 to the probe mobile.
- I'_n = power received at the home base station from other interference of non-CDMA origins. This term is included to accommodate all other possible interference sources that could be jamming the system in the CDMA band.
- N = thermal noise power.
- (W/R) = processing gain.

Note that if there is only one mobile being served by the base station (see Figure 7.5), there is nothing contributing to the reverse-link interference (except jammers and thermal noise). Therefore, (7.16) only has two interference terms I'_n and N in the denominator.

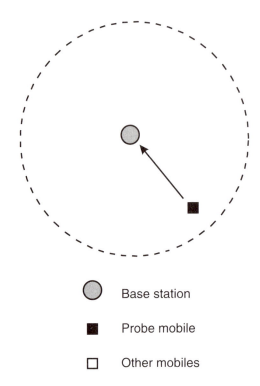

Base station

Probe mobile

Other mobiles

Figure 7.5 A single cell serving a single mobile—reverse-link characterization.

7.3.1.2 Single Cell and Many Mobiles

If there are many mobiles being served by one single cell (see Figure 7.6), then there will be an additional interference term. This additional interference term is due to the fact that mobiles (other than the probe mobile) are also transmitting on the reverse link. These additional powers on the reverse link are interfering with the reverse traffic channel of the probe mobile. The reverse link E_b/N_0 in this case is given by

$$\frac{E_b}{N_0} = \frac{T'L'_0(\theta_0,d_0)G_0(\theta_0)}{I'_m+I'_n+N}\left(\frac{W}{R}\right) \tag{7.17}$$

The additional term I'_m in the denominator is the total interference introduced by the reverse traffic channel transmissions of all the mobiles (other than the probe mobile or mobile 0). The value I'_m is given by

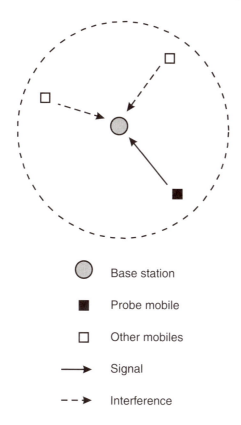

Base station

Probe mobile

Other mobiles

—————▶ Signal

– – –▶ Interference

Figure 7.6 A single cell serving many mobiles—reverse-link characterization.

$$I'_m = \sum_{j=1}^{J} T'_j L'_j \left(\theta_j, d_j \right) G_0 \left(\theta_j \right)$$ (7.18)

where T'_j is the reverse traffic channel ERP of mobile j. $L'_j(\theta_j, d_j)$ is the reverse path loss from mobile j in the direction θ_j back to the home base station a distance d_j away (i.e., the reverse path loss between mobile j and the home base station is a function of the mobile's direction θ_j and distance d_j). $G_0(\theta_j)$ is the receive antenna gain of the home base station (base station 0) in the direction θ_j to mobile j. Effectively, I'_m is the total reverse link interference introduced by mobiles served by the home base station.

Due to reverse-link power control, T'_j is dynamically changing in order to maintain an acceptable reverse-link quality for mobile j. Thus, T'_j is random, and consequently I'_m is also random. In practice, one may either assume a nominal value for I'_m or perform a Monte Carlo simulation to assess the simulated value of I'_m.

7.3.1.3 Many Cells and Many Mobiles

In the case where there are many cells and many mobiles (see Figure 7.7), there would be other mobiles (being served by other cells) that are interfering with the reverse traffic channel of the probe mobile. These other mobiles are taken into account by an additional interference term in the denominator. The reverse link E_b/N_0 in this case is given by

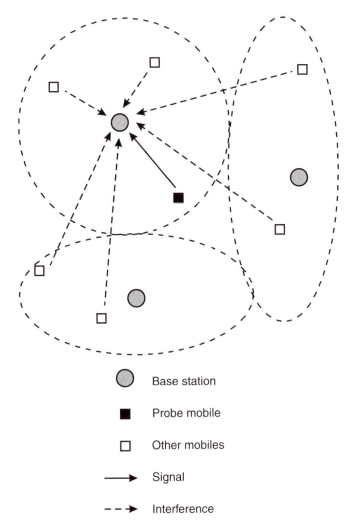

⬤	Base station
■	Probe mobile
☐	Other mobiles
⟶	Signal
- - ➤	Interference

Figure 7.7 Many cells serving many mobiles—reverse-link characterization.

$$\frac{E_b}{N_0} = \frac{T'L'_0(\theta_0, d_0)G_0(\theta_0)}{I'_m + I'_t + I'_n + N}\left(\frac{W}{R}\right) \tag{7.19}$$

The additional term I'_t in the denominator is the total interference introduced by the reverse traffic channel transmissions of all other mobiles that are not served by the home base station. Effectively, I'_t is the total reverse-link interference introduced by the mobiles served by other base stations (i.e., base stations other than the home base station). I'_t is given by

$$I'_t = \sum_{k=1}^{K} Y_k \tag{7.20}$$

where Y_k is the total traffic channel power (on the reverse link) received from those mobiles served by base station k. Note that the summation is from 1 to K, excluding the home base station 0, and K is the total number of base stations or sectors. We can arrive at Y_k by summing the received traffic channel powers from those mobiles served by base station k; that is, for base station k,

$$Y_k = \sum_{j=1}^{J_k} T'_{k,j} L'_{k,j}\left(\theta_{k,j}, d_{k,j}\right)G_0\left(\theta_{k,j}\right) \tag{7.21}$$

where for base station k, $T'_{k,j}$ is the reverse traffic channel ERP of mobile j. Again, for base station k, $L'_{k,j}(\theta_{k,j}, d_{k,j})$ is the reverse path loss from mobile j in the direction $\theta_{k,j}$ back to the home base station a distance $d_{k,j}$ away; that is, the reverse path loss between mobile j (served by base station k) and the home base station is a function of the mobile's direction $\theta_{k,j}$ and distance $d_{k,j}$. $G_0(\theta_{k,j})$ is the receive antenna gain of the home base station (base station 0) in the direction $\theta_{k,j}$ to mobile j (served by base station k).

7.3.2 Reverse-Link Rise

Having described all the reverse-link interference terms in the previous sections, we can define a useful factor in CDMA system design. The *rise R* of the reverse link is defined as the rise of the interference level above the thermal noise level; that is,

$$R = \frac{I'_m + I'_t + I'_n + N}{N} \tag{7.22}$$

The value R can be arrived at using Monte Carlo simulations, or it can be measured by determining the total power within the CDMA band and dividing the total power by the thermal noise power N. The measured total power is a close approximation of the numerator (i.e., $(I'_m + I'_t + I'_n + N)$). The value R is a very good indicator of whether or not the base station is heavily loaded on the reverse link. The higher the R, the greater the reverse-link load and the less likely the base station can support an additional user. Note that an increase in any one of the three interference terms, I'_m, I'_t, and I'_n, can cause R to go up.

7.3.3 Frequency Reuse Factor

In Chapter 4 we introduced the concept of the frequency reuse factor. We defined the frequency reuse factor F on the reverse link as

$$F = \frac{1}{1 + \eta} \tag{7.23}$$

where η is the loading factor. In terms of interference contributions, the loading factor η for the home base station can be defined as

$$\eta = \frac{I'_t}{I'_m} \tag{7.24}$$

where I'_t is the interference loading of traffic channels from mobiles that are in other cells, and I'_m is the interference loading of traffic channels from mobiles that are in the home cell. Substituting (7.24) into (7.23), we get

$$F = \frac{1}{1 + \dfrac{I'_t}{I'_m}} = \frac{I'_m}{I'_m + I'_t} \tag{7.25}$$

Remember that ideally, a CDMA system has a frequency reuse factor of 1 where the same physical RF band is being reused in every cell. However, cochannel RF interference from neighboring cells decreases the effectiveness of this frequency reuse scheme. Equation (7.25) characterizes that effectiveness in terms of interference loading of home cells and other cells. Note that if interference loading of other cells (i.e., I'_t) can be decreased, then F would increase. In the limit for a single-cell system, $I'_t = 0$ and thus $F = 1$.

7.4 PN Offset Planning

We have mentioned in Chapters 4 and 5 that on the forward link, a logical channel (i.e., pilot, paging, sync, or traffic channel) is separated from other logical channels by using different Walsh functions. We also notice that in addition to being spread by the Walsh function, a logical channel is further multiplied by the short PN sequence. In effect, every logical channel on the forward link is multiplied by the same short PN sequence assigned to that particular base station (or sector).

This multiplication by the short PN sequence is done to provide another layer of isolation between forward links of different base stations. For example, suppose that base station 1 transmits a traffic channel using Walsh function 21, and base station 2 transmits a traffic channel also using Walsh function 21. If these two base stations are next to each other, then mutual interference will occur.

In reality, base station 1's traffic channel is further multiplied by a short PN sequence, and base station 2's traffic channel is also further multiplied by a different short PN sequence. The multiplication by these two different short PN sequences ensures that all logical channels (i.e., pilot, paging, sync, and traffic channels) of one base station are separated from the logical channels of another base station. For this purpose, each base station (or sector) is assigned a different short PN sequence.

7.4.1 Short PN Sequences

Each short PN sequence is generated using a shift register with 15 delay elements (see Chapter 3). The length of such a PN sequence is about 2^{15}, or 32,768 chips. If we shift a PN sequence by one chip, then we have effectively generated a different PN sequence. Therefore, given that the PN sequence is 32,768 chips in length, we could theoretically generate and use about 32,768 different PN sequences, and we have 32,768 different PN sequences available to assign to different base stations! Given that we have this many different short PN sequences, PN planning would be unnecessary.

One problem with this simplistic view is that a difference of one chip between different PN sequences provides very little isolation in a mobile communications environment. Given that the transmission rate is 1.2288 Mcps, the duration of each chip is

$$\frac{1 \text{ sec}}{1.2288 \times 10^6 \text{chips}} = 0.81380 \times 10^{-6} \text{ sec} = 0.81380 \, \mu \text{sec}$$

A time duration of 0.81380 sec corresponds to a propagation distance of 244.14m; that is,

$$\left(0.81380 \times 10^{-6} \sec\right)\left(3 \times 10^8 \, \frac{\text{meters}}{\sec}\right) = 244.14 \text{ meters}$$

where 3×10^8 m/sec is the speed of light. Now suppose that there are two base stations: base station 1 and base station 2. The PN sequence of base station 1 differs from that of base station 2 by one chip. A mobile is 488m away from base station 1 and 244m away from base station 2 (see Figure 7.8).

A distance of 488m corresponds to a delay of two chips, while a distance of 244m corresponds to a delay of one chip. Therefore, base station 1's PN sequence arriving at the mobile would appear to the mobile as a PN sequence with a shift of two chips, and base station 2's PN sequence arriving at the mobile would appear as a PN sequence with a shift of one chip. These two received PN sequences cannot be distinguished from one another. In other

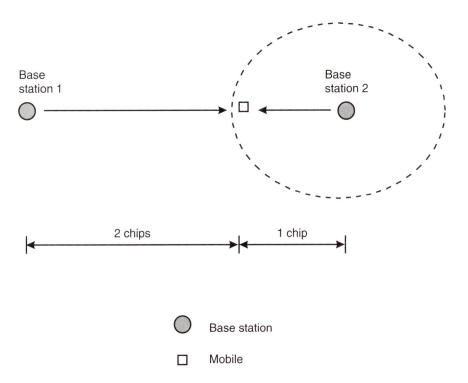

Figure 7.8 A situation where a mobile cannot distinguish the received PN sequences of two base stations.

words, the mobile would not know which PN sequence came from which base station (see Figure 7.9).

In order to provide more isolation among PN sequences that can be assigned (used), the IS-95 standard specifies that usable PN sequences need to have a minimum separation of 64 chips between each other. Each usable PN sequence is defined by its PN offset. For example, a PN sequence with PN offset 1 is different from a PN sequence with PN offset 0 by 64 chips, while a PN sequence with PN offset 4 is different from a PN sequence with PN offset 0 by $4 \times (64 \text{ chips}) = 256$ chips. By specifying a minimum separation of 64 chips, the total number of usable PN sequences becomes

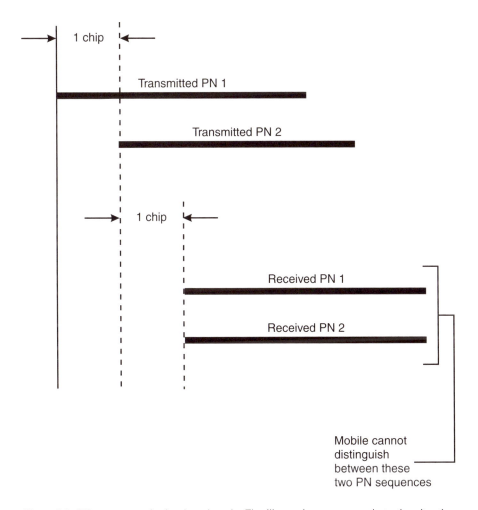

Figure 7.9 PN sequences in the time domain. The illustration corresponds to the situation shown in Figure 7.8.

$$\frac{32,768 \, \text{chips}}{64 \, \text{chips}} = 512$$

Therefore, given that there is a minimum separation of 64 chips, the maximum number of usable PN sequences is 512. The separation can be increased further by using the PILOT_INC parameter specified by the standard. If PILOT_INC = 1, then the minimum separation is 1×64 chips = 64 chips. If PILOT_INC = 2, then the minimum separation becomes 2×64 chips = 128 chips. Taking PILOT_INC into account, the total number of usable PN sequences is given by

$$\frac{32,768 \, \text{chips}}{\text{PILOT\_INC} \times (64 \, \text{chips})} \tag{7.26}$$

Suppose that PILOT_INC = 4; then, there are only 128 usable PN sequences available for assignment. With a limited number of usable PN sequences available, PN sequence planning now becomes analogous to AMPS frequency planning. The goal is to assign available PN offsets to different sectors such that there is a minimal confusion among the various received PN sequences at the mobile.

Note that because the pilot channel is effectively broadcasting the PN sequence of the base station, a base station's identifying PN sequence is also referred to as the base station's *pilot* or *pilot offset*.

7.4.2 Co-PN Offset

If two base stations are using the same PN sequence (i.e., PN sequences with identical PN offsets), what is the minimum required distance between the two base stations? Figure 7.10 illustrates the situation. The mobile is located at the edge of cell 2 (served by base station 2) and is homed on base station 2. The distance between the mobile and base station 2 is such that the propagation delay is Y chips. The distance between the mobile and base station 1 is such that the propagation delay is X chips. Base station 1 is identified by PN sequence 1, and base station 2 is identified by PN sequence 2. In this illustration, both PN sequences have the same PILOT_PN, or the same PN offset.

Figure 7.11 illustrates the PN sequences in the time domain. PN sequence 1 (PN 1) and PN sequence 2 (PN 2) are aligned when they are transmitted from their respective base stations. This is so because they have the same PN offset. However, PN sequence 1 undergoes delay X before being received by the mobile, and PN sequence 2 undergoes delay Y before being received by

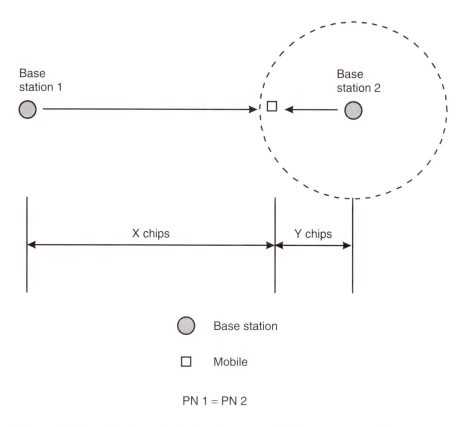

Base
station 1

Base
station 2

X chips

Y chips

⬤ Base station

☐ Mobile

PN 1 = PN 2

Figure 7.10 A situation illustrating PN offset planning of PN sequences with the same offset.

the mobile. The mobile uses search-window SRCH_WIN_A to intercept and receive pilot signals undergoing propagation delays. Note that SRCH_WIN_A is always centered on the earliest arriving pilot.

The mobile is currently being served by base station 2. As Figure 7.11 shows, if the received PN sequence 1 falls within SRCH_WIN_A of the mobile, the signal will be interpreted by the mobile to be a multipath of PN sequence 2. The mobile will then try to demodulate both pilot signals within SRCH_WIN_A and attempt to coherently combine them. As a result, interference occurs because these two signals don't have the same information contents. In this situation, PN sequence 1 is said to be an *alias* of PN sequence 2 (i.e., the mobile mistakenly thinks that received PN sequence 1 is a multipath component of PN sequence 2).

In order to avoid PN sequence aliasing, delay X has to be large enough for PN 1 to fall outside of SRCH_WIN_A. In other words, X must be larger

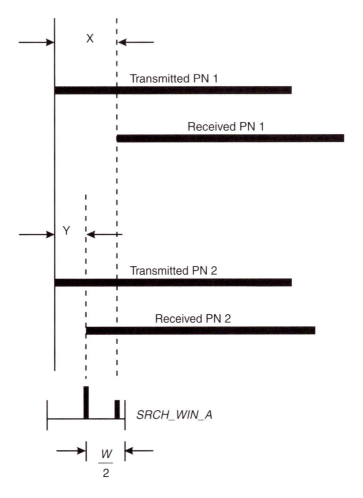

Figure 7.11 PN sequences in the time domain. The illustration corresponds to the situation shown in Figure 7.10.

than the sum of Y and $W/2$, where W is the window size of SRCH_WIN_A in chips. Therefore, the necessary condition to avoid PN offset aliasing is

$$X > Y + \frac{W}{2} \tag{7.27}$$

If we take Y as the coverage radius R (in chips) of cell 2 and $(X + Y) = D$ as the distance (in chips) between the two base stations, then we can manipulate (7.27) to

$$X - Y > \frac{W}{2}$$

$$X - Y + (X + Y) > \frac{W}{2} + (X + Y)$$

$$D > \frac{W}{2} + (X + Y) - X + Y$$

$$D > \frac{W}{2} + 2R$$

Since one chip corresponds to a distance of 244m, the condition of the physical distance d between two base stations that use the same PN offset is

$$d > 244\left(\frac{W}{2} + 2R\right)$$

or

$$d > 122W + 2r \tag{7.28}$$

where d is in meters, r is the coverage radius of base station 2 (home base station) in meters, and W is the size of SRCH_WIN_A in chips.

Note that separation in the time domain is not the only way to avoid PN offset aliasing. We can also use the received pilot strength to separate two pilots that have the same PN offset. If the path loss between base station 1 and the mobile is sufficiently large, then PN 1 would undergo high attenuation before reaching the mobile. Then, even if PN 1 does fall within SRCH_WIN_A, PN 1 would have a very low pilot strength, and the mobile would not be able to demodulate it.

7.4.3 Adjacent PN Offset

If two base stations are using adjacent PN sequences (i.e., PN sequences that are separated by (PILOT_INC $\times$ 64) chips), what is the necessary condition for no PN offset aliasing? Figure 7.12 illustrates the situation. This situation is identical to that described in Section 7.4.2 except that the two PN sequences are separated by $I = $ (PILOT_INC $\times$ 64) chips.

Figure 7.13 illustrates the PN sequences in the time domain. PN sequence 1 (PN 1) and PN sequence 2 (PN 2) are *not* aligned in time when they are transmitted from their respective base stations; instead, there is a shift

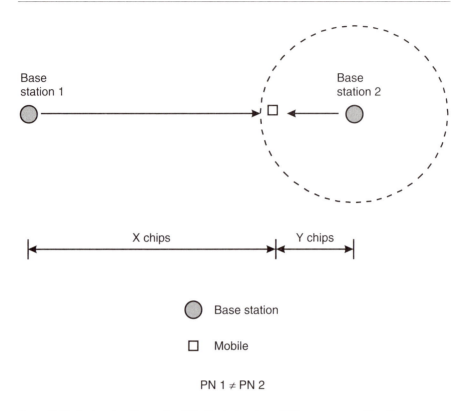

Figure 7.12 A situation illustrating PN offset planning of PN sequences with adjacent PN offsets.

of *I* chips between the two sequences. PN sequence 1 undergoes delay *X* before being received by the mobile, while PN sequence 2 undergoes delay *Y* before being received by the mobile.

Since the mobile is being served by base station 2, SRCH_WIN_A is normally centered on the received PN sequence 2 (assuming this pilot component is the earliest arriving multipath). As Figure 7.13 shows, if the received PN sequence 1 falls into SRCH_WIN_A of the mobile, the signal will be interpreted by the mobile to be a multipath of PN sequence 2. Since the aliased PN sequence 1 arrives seemingly before PN sequence 2, the mobile would shift its SRCH_WIN_A to center on PN sequence 1 and attempt to demodulate and combine both pilot signals. The result is interference, and usually a dropped call. Again, the mobile mistakenly thinks that received PN sequence 1 is an earlier arriving multipath component of PN sequence 2.

Figure 7.13 shows that in order to avoid PN offset aliasing, the following needs to be true:

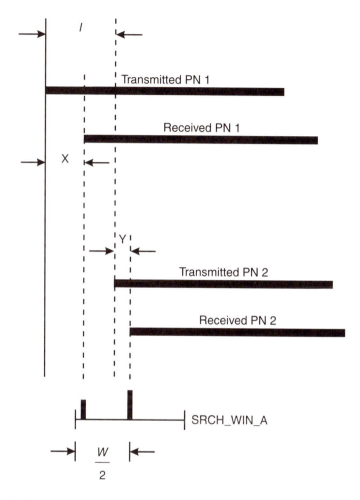

Figure 7.13 PN sequences in the time domain. The illustration corresponds to the situation shown in Figure 7.12.

$$X < I + Y - \frac{W}{2}$$

or

$$X + Y < I - \frac{W}{2} + 2Y$$

$$D < I - \frac{W}{2} + 2R \tag{7.29}$$

Since one chip corresponds to a distance of 244m, the condition of the physical distance d between two base stations that use adjacent PN offsets is

$$d < 244\left(I - \frac{W}{2} + 2R\right)$$

or

$$d < 244I - 122W + 2r \qquad (7.30)$$

where d is in meters, r is the coverage radius of base station 2 (home base station) in meters, and W is the size of SRCH_WIN_A in chips. Equation (7.30) is the condition for no aliasing between two base stations using adjacent PN sequences separated by I.

Equation (7.30) shows that the physical distance d has to be less than an upper bound $(244I - 122W + 2r)$. This condition is almost always satisfied with neighboring cells that are physically close to each other. In other words, (7.30) is most easily satisfied when we assign adjacent PN sequences to those cells that are physically close and similar in size [1].

7.5 9.6-Kbps and 14.4-Kbps Systems

In this section, we present a basic comparison between a system using 9.6-Kbps vocoders and a system using 14.4-Kbps vocoders. We outline the major differences between the two systems and use some performance estimates to approximate differences in coverage and capacity. It is important to recognize that the numerical values used are only rough estimates and are for discussion purposes only. Actual performance values depend on many factors and may differ significantly from system to system.

7.5.1 Voice Quality

The 14.4-Kbps vocoder is used primarily for its higher voice quality. Listening tests have shown that the perceived voice quality of a 9.6-Kbps vocoder at 1% FER is roughly equivalent to that of a 14.4 Kbps vocoder at 3% FER [2]. For the two vocoders, Figure 7.14 illustrates an approximate relationship between *mean opinion score* (MOS) and FER. As expected, voice quality as characterized by MOS decreases as FER increases. For a fixed FER, the 14.4-Kbps vocoder delivers a higher voice quality than what can be delivered by a 9.6-Kbps vocoder. For a given voice quality, the 14.4-Kbps vocoder is able to sustain a higher FER than what can be sustained by a 9.6-Kbps vocoder. Since lower FER typically requires a higher E_b/N_0, and since a higher required E_b/N_0 typically implies smaller coverage (or range), Figure 7.14 shows the classic trade-off between *quality* and *coverage*. Furthermore, since capacity is inversely proportional to the required E_b/N_0 (see Section 4.2), and since a higher required E_b/N_0 implies less capacity, Figure 7.14 shows another classic trade-off between *quality* and *capacity*. Namely, a higher voice quality demands that the system have less coverage and/or less capacity.

7.5.2 Power Control—Forward Link

The system using 14.4-Kbps vocoders and the system using 9.6-Kbps vocoders have very different forward power-control algorithms. In the 9.6-Kbps system, the mobile transmits *power measurement report messages* (PMRMs) over the traffic channel to inform the base station of the quality of the forward link. Based on the received PMRM, the base station adjusts the forward-link power

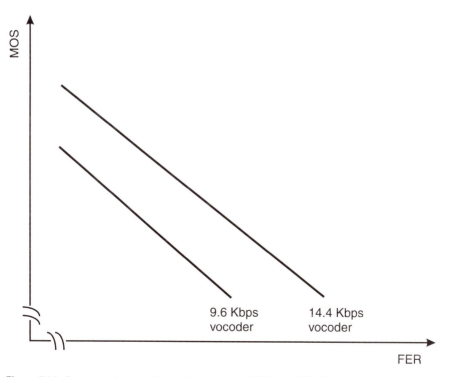

Figure 7.14 An approximate relationship between MOS and FER. The curves are for illustrative purposes only.

delivered to the mobile (see Section 4.3.3). The mobile can transmit at most four PMRMs every second.

In the 14.4-Kbps system, an *erasure indicator bit* is used in each reverse traffic frame to report the quality of a corresponding forward traffic frame. For each received frame on the forward link, the mobile determines whether a frame erasure has occurred. If the mobile determines that a forward-link frame is bad, then it sets the erasure indicator bit to 1 in the corresponding transmitted frame on the reverse link. If the mobile determines that a forward-link frame is good, then it sets the erasure indicator bit to 0 in the corresponding transmitted frame on the reverse link. Therefore, the mobile continuously monitors the quality of the forward link and continuously reports that quality (on the reverse link) once every frame. The mobile thus reports the forward-link quality once every 20 ms, or 50 times a second.

In the 14.4-Kbps system, this *fast power control* enables the mobile to report the forward-link condition more frequently. Thus, the power tracking performance is much better in the 14.4-Kbps system. The ultimate benefit of this enhanced tracking performance is reducing the required E_b/N_0 on the forward link.

7.5.3 Coverage

7.5.3.1 Forward Link

The primary difference between a 9.6-Kbps system and a 14.4-Kbps system is the processing gain. The processing gain for a 9.6-Kbps system is

$$\frac{W}{R} = \frac{1.2288 \times 10^6}{9.6 \times 10^3} = 128.00 = 21.072\,\text{dB}$$

The processing gain for a 14.4-Kbps system is

$$\frac{W}{R} = \frac{1.2288 \times 10^6}{14.4 \times 10^3} = 85.333 = 19.311\,\text{dB}$$

The difference in processing gain between the two systems is

$$21.072\,\text{dB} - 19.311\,\text{dB} = 1.76\,\text{dB}$$

This 1.76-dB difference in processing gain directly affects the link performance (see Section 7.2.2).

Furthermore, the 14.4-Kbps system uses a convolutional code of $R = 3/4$, while the 9.6-Kbps system uses a more powerful convolutional code of $R = 1/2$. This is done in order to accommodate a baseband rate of 14.4 Kbps (see Chapter 5). The result of this higher code rate (for the 14.4-Kbps system) is less error protection; thus, a higher E_b/N_0 is required on the forward link to achieve a specified FER. Various analyses have shown that in a mobile communications environment, the difference in the required E_b/N_0 between a 14.4-Kbps system and a 9.6-Kbps system is significant. The difference in required E_b/N_0 to achieve 1% FER varies between 2.2 and 5.3 dB [2].

However, we know that fast power control works to improve the forward link and reduce the required E_b/N_0. The reduction in required E_b/N_0 due to fast power control has been shown to vary between 1.4 and 4.9 dB [3]. Therefore, the gain afforded by fast power control may largely counteract the increased E_b/N_0 requirement in a 14.4-Kbps system.

Example 7.1

Previously, a base station operating in 9.6-Kbps mode could just barely serve a shopping mall 5 km away on the forward link. The base station used a maximum traffic channel ERP of 46 dBm. The RF engineer assumed a required E_b/N_0 of 8 dB in 9.6-Kbps mode. The base station now operates in 14.4-Kbps mode. Would it still be able to serve that shopping mall (assuming that other conditions, such as loading, have not changed)?

Assume that in 14.4-Kbps mode, the increase in E_b/N_0 requirement is 2.5 dB (due to the change in convolutional code rate). However, the required E_b/N_0 is reduced by 1.5 dB due to fast power control.

Solution

No. First of all, the 14.4-Kbps system requires an additional 1.0 dB in the required E_b/N_0 (i.e., 2.5 - 1.5 = 1.0 dB).

If the base station barely served the shopping mall before, then it must have operated at its maximum traffic channel ERP of 46 dBm most of the time. In 14.4-Kbps mode, the link first of all degrades by 1.76 dB due to the decrease in processing gain; thus, the base station needs to increase its maximum traffic channel ERP by at least 1.76 dB to overcome the processing gain loss. Second of all, the E_b/N_0 requirement has increased by 1.0 dB; thus, the base station needs to increase its maximum traffic channel ERP by at least another 1.0 dB to achieve the required E_b/N_0.

Therefore, the base station would not be able to serve the same shopping mall as before (i.e., its coverage area has shrunk). In order to serve the shopping mall, the RF engineer needs to increase the maximum traffic channel ERP by at least 2.76 dB (i.e., a maximum traffic channel ERP of 48.76 dBm at the base station).

7.5.3.2 Reverse Link

In 14.4-Kbps mode, the reverse link also suffers the same loss of 1.76 dB in processing gain. This 1.76-dB difference in processing gain directly affects the link performance (see Section 7.3).

In addition, on the reverse link, the 14.4-Kbps system uses a convolutional code of $R = 1/2$, while the 9.6-Kbps system uses a more powerful convolutional code of $R = 1/3$. This is done in order to accommodate a baseband rate of 14.4 Kbps (see Chapter 5). The result of this higher code rate (for the 14.4-Kbps system) is less error protection, and thus a higher E_b/N_0 is required on the link to achieve a specified FER. However, the increase in E_b/N_0 requirement is minimal. The difference in required E_b/N_0 to achieve 1% FER varies between 0.5 and 0.75 dB [2].

Example 7.2

Previously, an RF engineer could make a 9.6-Kbps CDMA call from the same shopping mall to the same base station (mentioned in Example 7.1). However, from the mall, her mobile phone consistently transmitted at a maximum ERP of 24 dBm. The base station now operates in 14.4-Kbps mode. Would she still be able to make a call from the same shopping mall on the reverse link?

We assume that other conditions, such as loading, have not changed. We also assume that in switching from 9.6-Kbps mode to 14.4-Kbps mode, the required E_b/N_0 has increased by 0.5 dB.

Solution

No. If the mobile phone transmitted at maximum ERP before, then it must have just barely closed the reverse link before. In 14.4-Kbps mode, the reverse link first of all degrades by 1.76 dB due to the decrease in processing gain; thus, the mobile needs to increase its maximum ERP by at least 1.76 dB to overcome the processing gain loss. Second of all, the E_b/N_0 requirement has increased by 0.5 dB; thus, the mobile needs to increase its maximum ERP by at least another 0.5 dB to achieve the required E_b/N_0.

Since the maximum ERP of the mobile cannot go beyond 24 dBm, the RF engineer would not be able to make a call in 14.4-Kbps mode (i.e., the reverse-link coverage has shrunk). In order to originate a call from the mall, the RF engineer needs to increase the maximum traffic channel ERP by at least

2.26 dB. Since she won't be able to do that, she could increase the receive antenna gain at the base station by at least 2.26 dB in order to close the reverse link.

7.5.4 Capacity

In Chapter 4, the system capacity (on the reverse link) was shown to be directly proportional to the processing gain and inversely proportional to the required E_b/N_0. It turns out the same relation also holds true for the forward link. We can calculate a relative capacity measure between a 14.4-Kbps system and a 9.6-Kbps system. Let $M_{14.4}/M_{9.6}$ be the relative capacity between the two systems:

$$\frac{M_{14.4}}{M_{9.6}} = \frac{\left(\dfrac{W/R_{14.4}}{\left(E_b/N_0\right)_{\text{required},14.4}}\right)}{\left(\dfrac{W/R_{9.6}}{\left(E_b/N_0\right)_{\text{required},9.6}}\right)} = \left(\frac{W/R_{14.4}}{W/R_{9.6}}\right)\left(\frac{\left(E_b/N_0\right)_{\text{required},9.6}}{\left(E_b/N_0\right)_{\text{required},14.4}}\right)$$

(7.31)

Example 7.3

What is the relative capacity on the forward link between a 14.4-Kbps system and a 9.6-Kbps system? Use the assumptions described in Example 7.1.

Solution

$$\frac{M_{14.4}}{M_{9.6}} = \left(-1.76\,\text{dB}\right) + \left(-1.0\,\text{dB}\right) = -2.76\,\text{dB} = 0.53 = 53\%$$

Example 7.4

What is the relative capacity on the reverse link between a 14.4-Kbps system and a 9.6-Kbps system? Use the assumptions described in Example 7.2.

Solution

$$\frac{M_{14.4}}{M_{9.6}} = \left(-1.76\,\text{dB}\right) + \left(-0.5\,\text{dB}\right) = -2.26\,\text{dB} = 0.59 = 59\%$$

References

[1]   Chang, C. R., J. Z. Wan, and M. F. Yee, "PN Offset Planning Strategies for Non-Uniform CDMA Networks," *Proc. 47th Annual Vehicular Technology Conf.*, IEEE, 1997.

[2]   Ketchum, J., M. Wallace, and R. Walton, " CDMA Network Deployment of 8 Kbps and 13 Kbps Voice Services," *Proc. of the International Conf. on Universal Personal Communications*, IEEE, 1996.

[3]   Stellakis, H., and R. Walton, "Forward Link Power Allocation," TIA TR45.5 contribution, Feb. 1995.

Select Bibliography

Aksu, A., and R. Walton, "14.4 Kbps Forward Link Performance," TIA TR45.5 contribution, Feb. 1995.

Sklar, B., *Digital Communications: Fundamentals and Applications*, Englewood Cliffs, NJ: Prentice Hall, 1988.

Viterbi, A. J., *CDMA Principles of Spread Spectrum Communication*, Reading, MA: Addison-Wesley, 1995.

Yang, J., D. Bao, and M. Ali, "PN Offset Planning in IS-95 Based CDMA Systems," *Proc. 47th Annual Vehicular Technology Conf.*, IEEE, 1997.

8

CDMA Performance Engineering

8.1 Introduction

Performance engineering refers to the process of optimizing the RF performance of a cell, a cluster of cells, or a system. While the design may look great on paper, the mobile communications environment is such that field engineering and optimization are almost always a necessity. This is especially so in CDMA because there are many parameters that can be adjusted, and the effects of parameter adjustments can be very pronounced.

In this chapter, we address several different topics of interest to an RF system engineer who is engineering and optimizing CDMA performance based on field observations. The chapter takes a modular approach to these different topics. Readers may find it convenient to read those sections that interest them.

8.2 Channel Supervision

8.2.1 Forward Link

There are many reasons why the mobile may drop a call. On the forward link, the mobile monitors each received traffic channel frame while in the traffic channel state. For each traffic channel frame, the mobile checks the frame quality indicator (CRC). If the frame quality indicator of a frame fails, or if the mobile is unable to decide on the data rate of the frame, then the mobile declares that the received frame is a *bad* frame [1]. Otherwise, the received frame is a *good* frame.

When in the traffic channel state, the mobile continuously monitors all the received traffic channel frames. If the mobile receives 12 (i.e., N_{2m} constant defined in IS-95A) consecutive bad frames, then it must shut off its transmitter.

However, if immediately after 12 consecutive bad frames are received, the mobile receives two (i.e., N_{3m} constant defined in IS-95A) consecutive good frames, then the mobile may again turn on its transmitter. Otherwise, the mobile loses the traffic channel, and a drop call occurs [1].

In addition, the mobile keeps a *fade timer* for the forward traffic channel. The timer is initialized when the mobile first turns on its transmitter while in the initialization substate of the traffic channel state; the fade timer is set to 5 (i.e., T_{5m} constant defined in IS-95A) sec and then runs down. This timer is reset to 5 sec every time the mobile receives two (i.e., N_{3m} constant defined in IS-95A) consecutive good frames. If the timer expires, then the mobile must shut off its transmitter and declare a loss of the forward traffic channel [1].

Furthermore, the mobile sometimes sends a message that requires an acknowledgment. When the mobile transmits a message that requires an acknowledgment, it waits for 0.4 (i.e., T_{1m} constant defined in IS-95A) sec. If it does not receive an acknowledgment within that time, the mobile retransmits the message and waits for another 0.4 sec. The mobile has three (i.e., N_{1m} constant defined in IS-95A) attempts to transmit a message that requires an acknowledgment. If the mobile does not receive an acknowledgment 0.4 sec after the third transmission, the mobile declares an acknowledgment failure.

8.2.2 Reverse Link

The same type of channel supervision also occurs on the reverse link. However, the IS-95 standard does not specifically specify what criteria to use to declare a loss of a traffic channel or an acknowledgment failure. Each infrastructure vendor is free to establish its own criteria.

8.3 Power-Control Parameters

The IS-95 standard specifies the power-control procedures on the forward link. The mobile reports to the base station the condition of the forward link. There are two modes of reporting: *periodic* and *threshold*. In the periodic reporting mode, the mobile reports frame error statistics at regular intervals; in the threshold reporting mode, the mobile only reports when frame errors reach a predetermined threshold.

There are five forward-link power-control parameters that are sent to the mobile in the overhead messages:

- PWR_THRESH_ENABLE—threshold report mode indicator;

- PWR_REP_THRESH—power-control reporting threshold;

- PWR_PERIOD_ENABLE—periodic report mode indicator;

- PWR_REP_FRAMES—power-control reporting frame count;

- PWR_REP_DELAY—power report delay.

The mobile uses its own counters, TOT_FRAMES and BAD_FRAMES, to keep track of the number of total frames and the number of bad frames received. Each time the mobile receives a frame, it increments the counter TOT_FRAMES by one. Each time the mobile receives a bad frame, it increments the counter BAD_FRAMES by one.

If PWR_THRESH_ENABLE = 1, then threshold reporting is enabled. When the BAD_FRAMES = PWR_REP_THRESH condition is met, the mobile reports frame error statistics by sending a PMRM to the base station [1]. After sending the PMRM, the mobile resets BAD_FRAMES (and TOT_FRAMES) to zero and will not increment BAD_FRAMES and TOT_FRAMES for a period of (PWR_REP_DELAY × 4) frames [1].

If PWR_PERIOD_ENABLE = 1, then periodic reporting is enabled. When the TOT_ FRAMES $= \left\lfloor 2^{(\text{PWR\_REP\_FRAMES}/2)} \times 5 \right\rfloor$ condition is met, the mobile reports frame error statistics by sending a PMRM to the base station. After sending the PMRM, the mobile resets TOT_FRAMES (and BAD_FRAMES) to zero and will not increment TOT_FRAMES and BAD_FRAMES for a period of (PWR_REP_DELAY × 4) frames [1].

The reported PMRM contains the following two fields:

- ERRORS_DETECTED—number of frame errors detected. The mobile sets this field to equal to its counter BAD_FRAMES.

- PWR_MEAS_FRAMES—number of forward traffic channel frames in the measurement period. The mobile sets this field to equal to its counter TOT_FRAMES.

The base station uses these two fields to calculate error statistics. Clearly, the mobile can be made to report PMRMs more often. This is done by adjusting the power-control parameters just described. For example, both reporting modes can be enabled simultaneously, or PWR_REP_THRESH and PWR_REP_FRAMES can be decreased. The trade-off is that more overhead messages occur on the reverse link due to more PMRMs.

8.4 Search-Window Sizes

The mobile uses the following three search windows to track the received pilot signals:

- SRCH_WIN_A—search-window size for the active and candidate sets;
- SRCH_WIN_N—search-window size for the neighbor set;
- SRCH_WIN_R—search-window size for the remaining set.

8.4.1 SRCH_WIN_A

SRCH_WIN_A is the search window that the mobile uses to track the active and candidate set pilots. The size of this window needs to be set according to the anticipated propagation environment. The window should be large enough to capture all usable multipaths of a base station's signal, and at the same time the window should be as small as possible in order to maximize searcher performance.

Figure 8.1 depicts the multipath situation. Remember that the search window is centered around the earliest arriving usable signal. The direct path (path A) travels 1 km to the mobile, while the multipath (path B) effectively travels 4 km before reaching the mobile. Since one chip corresponds to a propagation distance of 244.14m, the direct path travels a distance of

$$\frac{1,000 \text{ meters}}{244.14 \text{ meters/chip}} = 4.1 \text{ chips}$$

and the multipath travels a distance of

$$\frac{4,000 \text{ meters}}{244.14 \text{ meters/chip}} = 16.4 \text{ chips}$$

Therefore, the difference in distance traveled between the two paths is

$$16.4 \text{ chips} - 4.1 \text{ chips} = 12.3 \text{ chips}$$

Figure 8.1 also shows the search window (SRCH_WIN_A) at the mobile. Note that the direct path (path A) arrives the earliest and is thus at the center of the search window, while the multipath (path B) arrives 12.3 chips later. In

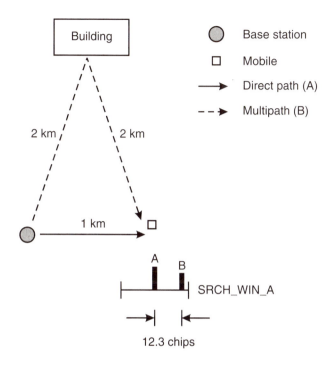

Figure 8.1 SRCH_WIN_A is used to capture the two paths A and B.

order for the search window to simultaneously capture these two paths, the window must be at least (2 × 12.3) chips, or 24.6 chips wide. In general, an RF engineer must set SRCH_WIN_A according to his or her knowledge of multi-path conditions within the cell. If there are many buildings, or if there is a large mountain in the distance that acts as a signal reflector, then the search-window size must be set accordingly.

Since large window sizes limit the searcher's performance, one should not set the search window size too large. Typically, longer multipath components travel greater distances and thus are more severely attenuated. By the time it reaches the mobile, the multipath may have too small of a signal strength to be usable. Therefore, a smaller SRCH_WIN_A can be used to limit the number of multipaths allowed.

Figure 8.2 depicts a situation where SRCH_WIN_A can be used to reduce the area where the mobile can conduct soft handoff. During soft hand-off between two cells, the mobile tracks two different pilots from two different base stations within the search window. Note that after the mobile identifies a known pilot, the mobile subtracts the PN offset from the received pilot phase so that only the propagation delay is left. In the example shown in Figure 8.2,

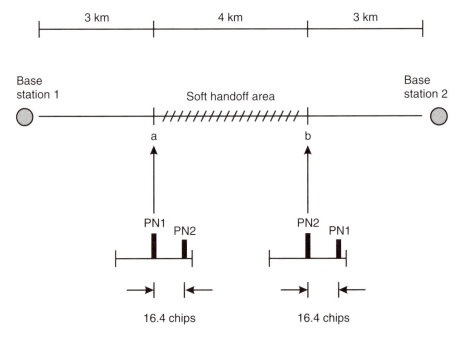

Figure 8.2 SRCH_WIN_A and a larger soft handoff area.

soft handoff currently occurs between points a and b. At point a, the mobile is 3 km from base station 1 and 7 km from base station 2. At point b, the mobile is 7 km from base station 1 and 3 km from base station 2. In terms of chips,

$$\text{At point a, the mobile is } \frac{3{,}000 \text{ meters}}{244.14 \text{ meters/chip}} = 12.3 \text{ chips from base station 1}$$

$$\frac{7{,}000 \text{ meters}}{244.14 \text{ meters/chip}} = 28.7 \text{ chips from base station 2}$$

$$\text{Path difference} = 28.7 \text{ chips} - 12.3 \text{ chips} = 16.4 \text{ chips}$$

$$\text{At point b, the mobile is } \frac{7{,}000 \text{ meters}}{244.14 \text{ meters/chip}} = 28.7 \text{ chips from base station 1}$$

$$\frac{3{,}000 \text{ meters}}{244.14 \text{ meters/chip}} = 12.3 \text{ chips from base station 2}$$

$$\text{Path difference} = 28.7 \text{ chips} - 12.3 \text{ chips} = 16.4 \text{ chips}$$

Figure 8.2 also shows the search windows at points a and b. At point a, the pilot from base station 2 lags the pilot from base station 1 by 16.4 chips. At point b, the pilot from base station 1 lags the pilot from base station 2 by 16.4 chips.

If an RF engineer wishes to contain the soft handoff area to between points c and d (shown in Figure 8.3), he or she would first repeat the same analysis. Now at point c, the mobile is 4 km from base station 1 and 6 km from base station 2. At point d, the mobile is 6 km from base station 1 and 4 km from base station 2. In terms of chips,

At point a, the mobile is $\dfrac{4{,}000 \text{ meters}}{244.14 \text{ meters/chip}} = 16.4$ chips from base station 1

$$\dfrac{6{,}000 \text{ meters}}{244.14 \text{ meters/chip}} = 24.6 \text{ chips from base station 2}$$

$$\text{Path difference} = 24.6 \text{ chips} - 16.4 \text{ chips} = 8.2 \text{ chips}$$

At point b, the mobile is $\dfrac{6{,}000 \text{ meters}}{244.14 \text{ meters/chip}} = 24.6$ chips from base station 1

$$\dfrac{4{,}000 \text{ meters}}{244.14 \text{ meters/chip}} = 16.4 \text{ chips from base station 2}$$

$$\text{Path difference} = 24.6 \text{ chips} - 16.4 \text{ chips} = 8.2 \text{ chips}$$

If the RF engineer wishes to contain the soft handoff area to a smaller area between points c and d, he or she should set SRCH_WIN_A to at least twice the *maximum* of the path differences (i.e., at least (2 × 8.2) chips, or 16.4 chips wide). This way, as the mobile travels from base station 1 to base station 2, the mobile can ensure that beyond point d, the pilot from base station 1 drops out of the search window.

8.4.2 SRCH_WIN_N and SRCH_WIN_R

SRCH_WIN_N is the search window that the mobile uses to track the neighbor set pilots. The size of this window is typically larger than that of SRCH_WIN_A. The window needs to be large enough not only to capture all usable multipaths of the home base station's signal, but also to capture the potential multipaths of neighbors' signals. Analysis methods discussed in the previous section can be used to determine SRCH_WIN_N. In this case, we

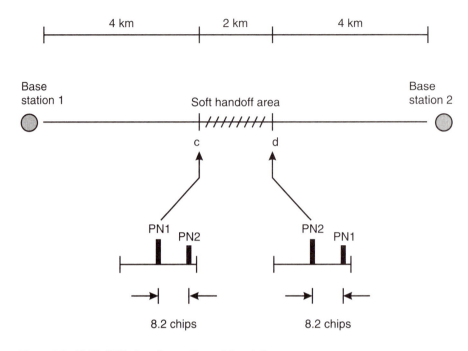

Figure 8.3 SRCH_WIN_A and a smaller soft handoff area.

need to take into account multipaths and path differences between the home base station and neighboring base stations.

We can derive an upper bound for SRCH_WIN_N. The maximum size of this search window is limited by the distance between two neighboring base stations. Figure 8.4 illustrates the concept. The mobile is located right next to base station 1, and therefore the propagation delay from base station 1 to the mobile is negligible. The distance between base station 1 (and the mobile) and base station 2 is 5 km. This distance in chips is

$$\frac{5{,}000 \text{ meters}}{244.14 \text{ meters/chip}} = 20.5 \text{ chips}$$

The search window shows that the pilot from base station 2 arrives 20.5 chips later. Thus, in order for a mobile (located anywhere within the cell) to search pilots of potential neighbors, SRCH_WIN_N needs to be set according to the physical distances between the current base station and its neighboring base stations. The actual size may not need to be this large, as this is an upper bound for SRCH_WIN_N.

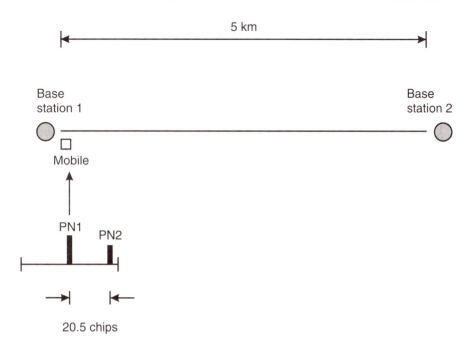

Figure 8.4 An upper bound for SRCH_WIN_N.

SRCH_WIN_R is the search window that the mobile uses to track the remaining set pilots. A typical requirement for the size of this window is that it is at least as large as SRCH_WIN_N.

8.5 Field Optimization

An RF engineer may drive around the system and measure actual field conditions. There are several typical problems associated with an unoptimized system. This section describes some of these conditions.

Before we start, it is important to note that field conditions change due to loading. An area may have excellent coverage at 10 P.M. when the traffic loading is light, but have poor coverage at 6 P.M. when the loading is heavy. Any field result would have to be examined in light of the loading condition at the time of measurement. Some rules-of-thumb apply, however. For example, if an area has poor coverage when the loading is light, the coverage would only get worse when the loading becomes heavy.

8.5.1 Pilot Strength

The pilot strength, or E_c/I_0, is a measure that reveals problem areas. The mobile requires sufficient E_c/I_0 to lock on or to remain on the system. A mobile may not even be able to originate in an area with excessively low E_c/I_0. E_c/I_0 is given by equation (7.7):

$$\frac{E_c}{I_0} = \frac{\alpha_0 P_0(\theta_0) L_0(\theta_0, d_0) G}{I_b + I_n + I_o + I_m + I_t + N}$$

As we can see from (7.7), low E_c/I_0 is typically caused by low pilot ERP (i.e., low $\alpha_0 P_0(\theta_0)$), excessive path loss (i.e., low $L_0(\theta_0, d_0)$), and/or high forward-link interference, which characterizes the denominator of (7.7).

Increasing the pilot ERP is one possible solution to low E_c/I_0. In the case of excessive path loss, adding an additional base station is another solution.

8.5.2 FER

FER is another measure that reveals problem areas. Because FER translates directly into perceived voice quality, the system must be optimized so that there is minimal and acceptable FER on both forward and reverse links. We examine FER in terms of the link E_b/N_0. On the forward link, E_b/N_0 is given by (7.15):

$$\frac{E_b}{N_0} = \frac{T_0(\theta_0) L_0(\theta_0, d_0) G}{I_b + I_n + I_o + I_{m'} + I_t + N} \left(\frac{W}{R} \right)$$

and on the reverse link, E_b/N_0 is given by (7.19):

$$\frac{E_b}{N_0} = \frac{T' L'_0(\theta_0, d_0) G_0(\theta_0)}{I'_m + I'_t + I'_n + N} \left(\frac{W}{R} \right)$$

An area with high FER indicates that E_b/N_0 has decreased below a certain threshold. We now examine some commonly observed causes for high FER in terms of (7.15) and (7.19).

8.5.3 Forward Link Coverage

One common cause of high forward-link FER is poor forward-link coverage. In this case, an area is characterized by high forward-link FER and dropped calls.

Poor forward-link coverage causes a low received E_b/N_0 on the forward link. By definition, poor forward-link coverage is due to excessive forward path loss (i.e., low $L_0(\theta_0,d_0)$ in (7.15)); poor forward-link coverage can also be due to low traffic channel ERP (i.e., low $T_0(\theta_0)$ in (7.15)).

The symptoms of poor forward-link coverage are *high forward FER* and *low mobile received power*. If excessive path loss is the cause, low measured E_c/I_0 can also be another manifestation. Because the mobile is consistently receiving high FER, there can also be an increased activity in PMRM transmissions.

One possible solution to poor forward-link coverage is to increase the maximum traffic channel ERP so that the base station is allowed to deliver higher powers on the traffic channel. However, this solution typically carries with it capacity implications. As one increases the forward traffic channel ERP, one also increases the amount of forward-link interference to those mobiles in the same cell, as well as to those mobiles in neighboring cells. Therefore, increasing traffic channel ERP has to be balanced with maintaining acceptable interference introduced to other mobiles. Another possible solution to poor forward-link coverage is to add an additional base station.

8.5.4 Forward-Link Interference

Another cause for high forward-link FER is high forward-link interference. High forward-link interference causes a low received E_b/N_0 on the forward link. In this case, the low E_b/N_0 is due to high interference terms in the denominator of (7.15).

The manifestations of high forward-link interference are *high forward FER* and *high mobile received power*. Because the mobile received power is a measurement of the total power in the receive band, high mobile received power and high forward FER are indicative of a lot of forward-link interference.

There are four typical sources of forward link interference, as follows.

The first source is interference power due to forward traffic channel transmission of the home base station. This interference is equivalent to the $I_{m'}$ term in the denominator of (7.15). This term is the sum of all (interference) powers due to traffic channel powers delivered to those mobiles in the same cell. One solution to reducing this type of interference is to limit the physical number of traffic channels (equipped) in the cell.

The second source is interference power due to forward overhead transmissions of other base stations. This interference is equivalent to the I_o term in the denominator of (7.15); this type of interference is sometimes known as the *pilot pollution* (due to the fact that pilot powers are typically the highest among all overhead channels). The I_o term is the sum of all overhead powers from neighboring base stations; thus, reducing overhead powers of neighboring base stations is one way to reduce this type of interference.

The third source is interference power due to forward traffic channel transmissions of other base stations. This interference is equivalent to the I_t term in the denominator of (7.15). This term is the sum of all (interference) powers due to traffic channel powers delivered by other base stations to other mobiles in other cells. One solution to reducing this interference is to examine the antenna orientation of neighboring base stations. By reorienting antennas of neighboring base stations, one may be able to reduce forward interference into the home cell. Of course, changes in antenna configuration need to be done without affecting the coverage of neighboring cells.

The fourth source is interference of non-CDMA origin. This interference corresponds to the I_n term in the denominator of (7.15). This term describes interference from non-CDMA transmitters such as jammers. Jammers often operate (intentionally or unintentionally) illegally in the CDMA band, and eliminating this interference requires identifying the interferer's location. This is often done by surveying the interferer's power and triangulating the position.

8.5.5　Reverse-Link Coverage

One common cause of high reverse-link FER is poor reverse-link coverage. An area with poor reverse-link coverage is characterized by high reverse-link FER and dropped calls. Poor reverse-link coverage causes a low received E_b/N_0 on the reverse link. By definition, poor reverse-link coverage is due to excessive reverse path loss (i.e., low $L'_0(\theta_0,d_0)$ in (7.19)).

Poor reverse-link coverage is characterized by *high reverse-link FER* and *high mobile transmit power*. This is because, due to reverse-link power control, the mobile attempts to close the reverse link by increasing its transmit power. Unfortunately, one cannot increase the maximum mobile transmit power to combat excessive path loss. The reason is that the mobile typically has a maximum transmit power limited by its power amplifier. One solution to poor reverse-link coverage is to add an additional base station.

8.5.6　Reverse-Link Interference

Another cause of high reverse-link FER is high reverse-link interference. High reverse-link interference causes a low received E_b/N_0 on the reverse link; the low E_b/N_0 is typically due to high interference terms in the denominator of (7.19).

The manifestations of high reverse-link interference are *high reverse FER* and/or *high base station received power*. Because the base station received power is a measurement of the total power in the receive band, high base station received power with high reverse FER is indicative of a lot of reverse-link interference.

There are three typical sources of reverse link interference, as follows.

The first source is interference power due to traffic channel transmissions of those mobiles in the same cell. This interference is equivalent to the I'_m term in the denominator of (7.19). This term is the sum of all (interference) powers due to traffic channel transmissions of mobiles in the same cell. One way to reduce I'_m is to reduce the home cell footprint (via downtilting or antenna change); by reducing the coverage footprint of the home cell, one can reduce the number of mobiles served by the home cell and thus reduce I'_m.

The second source is interference power due to traffic channel transmissions of those mobiles in other cells. This interference is equivalent to the I'_t term in the denominator of (7.19). This term characterizes the sum of all (interference) powers due to traffic channel transmissions of other mobiles in other cells. One solution to reducing this type of interference is to change the antenna orientation/downtilt of the home base station. By controlling the receive coverage footprint of the home base station, one may be able to limit the amount of interference power received from other mobiles (of other cells). Of course, the antenna change has to be done without adversely affecting the basic coverage of the home base station.

The third source is interference of non-CDMA origin. This interference corresponds to the I'_n term in the denominator of (7.19). This term describes interference from non-CDMA transmitters such as jammers. Jammers often operate (intentionally or unintentionally) illegally in the CDMA band, and eliminating this interference requires identifying the interferer's location. This is often done by surveying the interferer's power and triangulating the position.

8.5.7 Some Concluding Remarks

In this section, we have segmented CDMA performance problems into four causes, which are poor coverage and high interference on forward and reverse links. It is important to recognize that in reality, observed performance problems are often due to a combination of these underlying causes. For example, a combination of poor forward coverage and high forward interference typifies a problem commonly known as *no dominant server*. This problem is characterized by many pilots present in the affected area, but none of the pilots has an adequate E_c/I_0 to be a dominant server; because there are many pilots on the forward link, the resulting forward-link interference is high. The solutions to this situation include increasing the pilot ERP and increasing the forward traffic channel ERP of one base station so a dominant server can be forced into the problem area. Another solution is to add an additional base station.

Adjusting the search-window size is also an important optimization technique. Those pilots that are not captured by the active search window cannot be diversity-combined and thus become interference to the mobile.

If a cell's performance is limited by either forward or reverse interference, typically the capacity of that cell has been reached. To the extent that the interference can be reduced, the performance of that cell can be increased. It is important to keep in mind that managing CDMA performance is effectively managing interference and noise. We shall focus on the topic of noise management in the next chapter.

References

[1]   TIA/EIA IS-95A, "Mobile Station-Base Station Compatibility Standard for Dual-Mode Wideband Spread Spectrum Cellular System," Telecommunications Industry Association.

Select Bibliography

Garg, V. K., K. Smolik, and J. E. Wilkes, *Applications of CDMA in Wireless/Personal Communications*, Upper Saddle River, NJ: Prentice Hall, 1997.

Harte, L., *CDMA IS-95 for Cellular and PCS: Technology, Applications and Resource Guide*, New York, NY: McGraw-Hill, 1997.

Ketchum, J., M. Wallace, and R. Walton, "CDMA Network Deployment of 8 Kbps and 13 Kbps Voice Services," *Proc. of the International Conf. on Universal Personal Communications*, IEEE, 1996.

Sklar, B., *Digital Communications: Fundamentals and Applications*, Englewood Cliffs, NJ: Prentice Hall, 1988.

9

System Noise Management

9.1 Introduction

CDMA is a multiple access technology that utilizes direct-sequence spread-spectrum techniques. With this technology comes a paradigm shift. Contrary to the conventional FDMA and TDMA systems where noise rejection deals primarily with *out-of-band* noise, a CDMA system concerns mostly with *in-band* noise. This noise may come from self-jamming (or self-noise), background noise, man-made noise, intermodulation, or noise generated in the receiver. If one can reduce the unwanted in-band noise, such reduction translates directly into improved performance. Thus, the goal of an RF system engineer is to design a network that minimizes the amount of unwanted noise introduced into the receiver on both forward and reverse links.

Undesired noise comes from many different sources. This noise may come from natural or human sources. Naturally occurring noise includes atmospheric disturbances, background noise, and thermal noise generated in the receiver itself. Human interference comes from other communication and electrical systems that introduce, either intentionally or unintentionally, interference into the CDMA band. Through careful engineering, the effects of many unwanted signals can be reduced [1]. This chapter describes some of these interference sources and techniques to minimize them. Because "one man's music is another man's noise," we use the terms *noise* and *interference* interchangeably in this chapter. Oftentimes, noise refers to an unwanted signal of natural origin, while interference refers to unwanted signals of human origin. In Sections 9.2.1 and 9.2.2, we briefly review those interference sources that affect the performance (i.e., E_b/N_0) of forward and reverse traffic channels.

9.2 Types of Interference

9.2.1 Forward Link

Equation (7.15), reprinted below from Chapter 7, is an expression of E_b/N_0 for one single traffic channel on the forward link. By examining (7.15), we can see that there are six interference terms present in the denominator, and the magnitudes of these interference terms have direct impacts on E_b/N_0, and ultimately on the voice quality. Each of these terms takes into account interference powers generated by a particular source. The denominator of (7.15) takes into account those interference terms present on the CDMA forward link:

$$\frac{E_b}{N_0} = \frac{T_0(\theta_0)L_0(\theta_0,d_0)G}{I_b + I_n + I_o + I_{m'} + I_t + N}\left(\frac{W}{R}\right)$$

The first term is *interference power due to transmission of overhead channels of the same base station.* This interference is equivalent to the I_b term in the denominator of (7.15). Because the overhead channels (i.e., pilot, paging, and sync channels) are transmitted in the same band as the traffic channel, the transmission of these overhead channels represents in-band interference to the traffic channel. A formal expression of I_b is given by (7.2) in Chapter 7.

The second term is *interference power due to transmission of overhead channels of other base stations.* This interference is equivalent to the I_o term in the denominator of (7.15). Because other neighboring base stations are also transmitting their own overhead channels in the same band, the transmission of these overhead channels represents in-band interference to the traffic channel. A formal expression of I_o is given by (7.4).

The third term is *interference power due to forward traffic channel transmission of the same base station.* This interference is equivalent to the $I_{m'}$ term in the denominator of (7.15). This term is the sum of all (interference) traffic channel powers delivered to those mobiles in the same cell. Since these other traffic channels (in the same cell) are transmitted in the same band, they become interference to the E_b/N_0 of the desired traffic channel. A formal expression of I_m is given by (7.6).

The fourth term is *interference power due to forward traffic channel transmissions of other base stations.* This interference is equivalent to the I_t term in the denominator of (7.15). This term is the sum of all (interference) traffic channel powers delivered by other base stations to other mobiles in other cells. Again, since these other traffic channels (of neighboring cells) are transmitted in the same band, they become interference to the E_b/N_0 of the desired traffic channel. A formal expression of I_t is given by (7.8).

The fifth term is *interference of non-CDMA origin*. This interference corresponds to the I_n term in the denominator of (7.15). This term describes the (in-band) human interference from those communication and electrical systems that are not part of the CDMA system. These sources of interference are often called *jammers*.

The sixth term is *thermal noise*. This corresponds to the N term in the denominator of (7.15). These are thermal noises generated in the receiver due to the device's physical temperature. This type of noise is unavoidable in all communication systems.

9.2.2 Reverse Link

In the same way, (7.19), reprinted below from Chapter 7, is an expression of E_b/N_0 for a traffic channel on the reverse link. There are four interference terms present in the denominator. Each of these terms takes into account interference powers generated by a particular source. The denominator of (7.19) takes into account those interference terms present on the CDMA reverse link:

$$\frac{E_b}{N_0} = \frac{T'L'_0(\theta_0, d_0)G_0(\theta_0)}{I'_m + I'_t + I'_n + N}\left(\frac{W}{R}\right)$$

The first term is *interference power due to traffic channel transmissions of mobiles in the same cell*. This interference is equivalent to the I'_m term in the denominator of (7.19). This term is the sum of all (interference) powers due to traffic channel transmissions of those mobiles in the same cell. Because all traffic channels are transmitted in-band, these transmissions represent in-band interference to the E_b/N_0 of the desired traffic channel. A formal expression of I'_m is given by (7.18).

The second term is *interference power due to traffic channel transmissions of mobiles in other cells*. This interference is equivalent to the I'_t term in the denominator of (7.19). This term characterizes the sum of all (interference) powers due to traffic channel transmissions from other mobiles in other cells. A formal expression of I'_t is given by (7.20).

The third term is *interference of non-CDMA origin*. This interference corresponds to the I'_n term in the denominator of (7.19). This term describes the interference from non-CDMA transmitters such as jammers. These jammers are those who operate (intentionally or unintentionally) illegally in the CDMA band.

The fourth term is *thermal noise*. This corresponds to the N term in the denominator of (7.19). These are thermal noises generated in the receiver due to the device's physical temperature. This type of noise is present in all communication systems.

As we can see, one unavoidable source of noise is thermal noise N; the cause of which is the thermal motion of electrons in conductors. Since all communication systems have conductors in them, thermal noise is always present as well [1]. It is this thermal noise that presents a fundamental limit to system performance. We examine thermal noise in more detail in the next section.

9.3 Thermal Noise

Thermal noise is caused by the random motion of electrons in a conductor that has a physical temperature. From kinetic theory of particles, a particle at an absolute temperature T has an average energy that is proportional to kT, where k is the Boltzmann's constant. We state here without proof that a thermal noise source can be treated as a random noise generator with an open-circuit mean-square voltage $\bar{e}_n^2$ of

$$\bar{e}_n^2 = 4kTWR_i \tag{9.1}$$

where

- K = Boltzmann's constant (1.38×10^{-23} W/K/Hz or -228.6 dBW/K/Hz);
- T = temperature in kelvin;
- W = bandwidth in hertz;
- R_i = internal resistance of the noise generator in ohms.

We invoke the maximum power transfer theorem. This theorem states that maximum power is delivered to a load when the load impedance is equal to the complex conjugate of the generator impedance. Therefore, if the load impedance is equal to the internal resistance of the noise generator R_i, maximum noise power is transferred to the load. As shown in Figure 9.1, given that $R_l = R_i$, the voltage across the load is equal to half of the mean-square voltage of the noise generator [1]. Thus, the noise power transferred to the load is

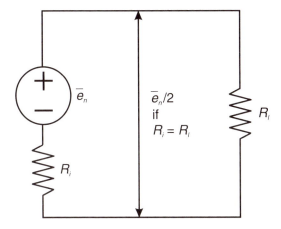

Figure 9.1 Equivalent circuit of thermal noise generator. After [1].

$$N = \frac{\left(\dfrac{\sqrt{4\,KTBR_i}}{2}\right)^2}{R_l} = kTW \tag{9.2}$$

Equation (9.2) is useful in describing the noise power in terms of a physical temperature of a communication device in question. As shown in (9.2), the higher the temperature, the higher the noise power. The noise power is also proportional to the bandwidth of the device. Given that the noise power density is kT W/Hz, the wider the bandwidth, the more noise is "let" in, and thus the higher the noise power. Readers are referred to [1] for more details on noise temperatures and noise figures.

9.4 Low-Noise Amplifier

In this section, we examine a technique to reduce the system noise figure on the reverse link. This reduction is accomplished by installing *low-noise amplifiers* (LNAs) at a CDMA base station. First of all, we derive signal-to-noise ratio (equivalent to E_b/N_0) expressions for two systems: a baseline system without LNAs and a system with LNAs. Then, we divide the two SNR expressions to obtain an expression for signal-to-noise ratio *improvement* obtained by using LNAs.

9.4.1 Baseline System Without LNAs

Figure 9.2 depicts a CDMA base station receiving system. The antenna has an effective antenna temperature T_A. The ambient temperature T_0 is 290K. The line loss between the antenna and the radio is L_1. The front end of the radio has a gain G_2 and a noise figure F_2. We assume that the signal right after the antenna has a signal strength of S_{in}, and we define the SNR at the predetection output to be $(S/N)_{out}$.

The CDMA base station receiving system consists of the antenna and the receiving equipment. The receiving equipment refers to those components behind the antenna. In our case, the receiving equipment contains the cable and the radio front end. The composite noise figure of the receiving equipment is

$$F_{comp} = F_1 + \frac{F_2 - 1}{G_1} \tag{9.3}$$

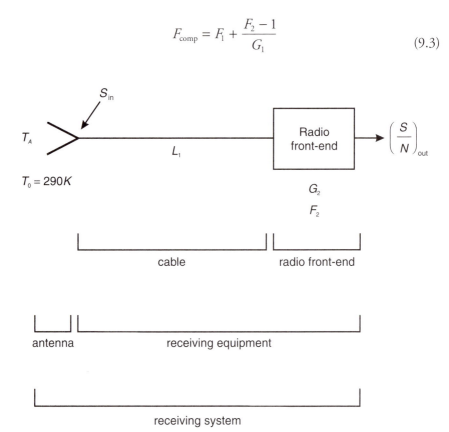

Figure 9.2 A CDMA base station with no LNAs.

The gain G_1 of the cable is simply the inverse of its loss L_1. If we assume that the noise figure of the cable is equal to its line loss (i.e., $F_1 = L_1$), then

$$F_{comp} = L_1 + L_1(F_2 - 1) \qquad (9.4)$$

For any receiving equipment, the composite noise temperature is

$$T_{comp} = (F_{comp} - 1)T_0 \qquad (9.5)$$

and the system noise temperature is the sum of the antenna temperature and the composite noise temperature; that is,

$$T_{sys} = T_A + T_{comp} \qquad (9.6)$$

If the antenna is aboard a satellite facing the earth, then the antenna temperature T_A would be 290K (i.e., ambient temperature) since the earth's temperature is effectively at 290K. However, in our case, the effective antenna temperature would be higher than the ambient temperature because the antenna is also receiving interference from other sources. We can use the definition of reverse-link rise to quantify this increase in interference. Recall in Chapter 7, (7.22), that we define the reverse link rise to be

$$R = \frac{I'_m + I'_t + I'_n + N}{N}$$

Since the actual noise seen by the antenna is effectively the numerator of (7.22), the actual noise can be written as

$$I'_m + I'_t + I'_n + N = RN = R(kT_0W) = k(RT_0)W = kT_AW \qquad (9.7)$$

Therefore, the effective antenna temperature (T_A) is equivalent to the rise multiplied by the ambient temperature (RT_0).

We now proceed to write an expression for the signal-to-noise ratio $(S/N)_{out}$ at the predetection output. For the numerator, the signal (S_{in}) right after the antenna undergoes a loss of $/L_1$ and a gain of G_2, while the system noise $(kT_{sys}W)$ also undergoes a loss of L_1 and a gain of G_2. The signal-to-noise ratio becomes

$$\left(\frac{S}{N}\right)_{\text{out}} = \frac{S_{\text{in}}(1/L_1)G_2}{(1/L_1)G_2\left(kT_{\text{sys}}W\right)} = \frac{S_{\text{in}}(1/L_1)G_2}{(1/L_1)G_2k\left(T_A + T_{\text{comp}}\right)W}$$

$$= \frac{S_{\text{in}}}{k\left(T_A + T_{\text{comp}}\right)W} \tag{9.8}$$

We already know that $T_A = RT_0$. For T_{comp}, we substitute (9.4) into (9.5) and get

$$T_{\text{comp}} = \left(F_{\text{comp}} - 1\right)T_0 = \left[L_1 + L_1(F_2 - 1) - 1\right]T_0 = (L_1F_2 - 1)T_0 \tag{9.9}$$

Substituting (9.9) into (9.8) yields

$$\left(\frac{S}{N}\right)_{\text{out}} = \frac{S_{\text{in}}}{k\left(T_A + T_{\text{comp}}\right)W} = \frac{S_{\text{in}}}{k\left(RT_0 + (L_1F_2 - 1)T_0\right)W}$$

$$= \frac{S_{\text{in}}}{kT_0\left(R + (L_1F_2 - 1)\right)W} \tag{9.10}$$

We can write (9.10) as

$$\left(\frac{S}{N}\right)_{\text{out}} = \frac{S_{\text{in}}}{k\sigma T_0 W} \tag{9.11}$$

where σ is the *noise enhancement* given by

$$\sigma = R + \left(L_1F_2 - 1\right) \tag{9.12}$$

9.4.2 System With LNA

Figure 9.3 shows a CDMA base station receiving system with an LNA. As before, the antenna has an effective antenna temperature T_A. The ambient temperature T_0 is 290K. However, an LNA is now placed between the antenna and the radio. The line loss between the antenna and the LNA is L'_1. The LNA itself has a gain G'_2 and a noise figure F'_2. The line loss between the LNA and the radio is L'_3, and the front end of the radio has a gain G'_4 and a noise figure F'_4. The low-noise amplifier is so called because its noise figure F'_2 is typically less than the noise figure F'_4 of the radio.

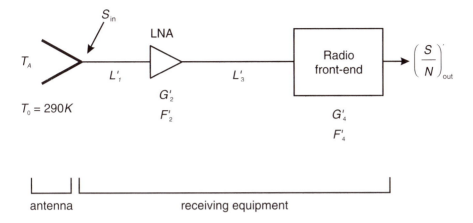

Figure 9.3 A CDMA base station with LNA.

We assume that the signal right after the antenna has a signal strength of S'_{in}, and we define the signal-to-noise ratio at the predetection output to be $(S/N)'_{out}$. In this case, the composite noise figure of the receiving equipment is

$$F'_{comp} = F'_1 + \frac{F'_2 - 1}{G'_1} + \frac{F'_3 - 1}{G'_1 G'_2} + \frac{F'_4 - 1}{G'_1 G'_2 G'_3} \qquad (9.13)$$

Equation (9.13) can be written as follows, recognizing that the gain of a cable is simply the inverse of its loss and assuming that the noise figure of the cable is equal to its line loss:

$$F'_{comp} = L'_1 + L'_1 (F'_2 - 1) + \frac{L'_1 (L'_3 - 1)}{G'_2} + \frac{L'_1 L'_3 (F'_4 - 1)}{G'_2} \qquad (9.14)$$

The system noise temperature is still the same as in (9.6):

$$T_{sys} = T_A + T_{comp}$$

We proceed to write an expression for the signal-to-noise ratio $(S/N)'_{out}$ at the predetection output. For the numerator, the signal (S'_{in}) right after the antenna undergoes a loss of L'_1, a gain of G'_2, a loss of L'_3, and a gain of G'_4. The system noise $(kT_{sys}W)$ also undergoes a loss of L'_1, a gain of G'_2, a loss of L'_3, and a gain of G'_4. The signal-to-noise ratio is thus

$$\left(\frac{S}{N}\right)'_{\text{out}} = \frac{S_{\text{in}}\left(1/L'_1\right)G'_2\left(1/L'_3\right)G'_4}{\left(1/L'_1\right)G'_2\left(1/L'_3\right)G'_4\left(kT_{\text{sys}}W\right)} = \frac{S_{\text{in}}}{kT_{\text{sys}}W} \tag{9.15}$$

By using (9.6) and (9.7), the above equation can be written as

$$\left(\frac{S}{N}\right)'_{\text{out}} = \frac{S_{\text{in}}}{k\left(RT_0 + T'_{\text{comp}}\right)W} \tag{9.16}$$

For T_{comp}, we substitute (9.13) into (9.5) and get

$$T'_{\text{comp}} = \left(F'_{\text{comp}}-1\right)T_0 = \left[L'_1 + L'_1\left(F'_2-1\right)\right.$$

$$\left. + \frac{L'_1\left(L'_3-1\right)}{G'_2} + \frac{L'_1 L'_3\left(F'_4-1\right)}{G'_2} - 1\right]T_0$$

or

$$T'_{\text{comp}} = \left(F'_{\text{comp}}-1\right)T_0 = \left[L'_1 F'_2 + \frac{L'_1\left(L'_3-1\right)}{G'_2} + \frac{L'_1 L'_3\left(F'_4-1\right)}{G'_2} - 1\right]T_0 \tag{9.17}$$

Substituting (9.17) into (9.16) yields

$$\left(\frac{S}{N}\right)'_{\text{out}} = \frac{S'_{\text{in}}}{k\left[R + L'_1 F'_2 + \dfrac{L'_1\left(L'_3-1\right)}{G'_2} + \dfrac{L'_1 L'_3\left(F'_4-1\right)}{G'_2} - 1\right]T_0 W} \tag{9.18}$$

We can write (9.18) as

$$\left(\frac{S}{N}\right)'_{\text{out}} = \frac{S_{\text{in}}}{k\sigma' T_0 W} \tag{9.19}$$

where σ' is the *noise enhancement* given by

$$\sigma' = R + L'_1 F'_2 + \frac{L'_1 (L'_3 - 1)}{G'_2} + \frac{L'_1 L'_3 (F'_4 - 1)}{G'_2} - 1 \tag{9.20}$$

Note that if G'_2 is large, then

$$\sigma' \approx R + L'_1 F'_2 - 1 \tag{9.21}$$

9.4.3 Signal-to-Noise Ratio Improvement

In the previous two sections, we derived an SNR expression for a system without LNAs (i.e., $(S/N)_{out}$), and an SNR expression for a system with LNAs (i.e., $(S/N)'_{out}$). We can divide these two expressions to get an expression for *SNR improvement Q*. The SNR improvement quantifies how much the SNR has improved by adding an LNA between the antenna and the radio. The SNR improvement is defined as

$$Q \equiv \frac{(S/N)'_{out}}{(S/N)_{out}} \tag{9.22}$$

and it can be evaluated as

$$Q = \frac{\left(\dfrac{S'_{in}}{k\sigma' T_0 W} \right)}{\left(\dfrac{S_{in}}{k\sigma T_0 W} \right)} = \frac{\sigma}{\sigma'} = \frac{(R-1) + L_1 F_2}{(R-1) + L'_1 F'_2 + \dfrac{L'_1 (L'_3 - 1)}{G'_2} + \dfrac{L'_1 L'_3 (F'_4 - 1)}{G'_2}} \tag{9.23}$$

If G'_2 is large, then (9.23) can be approximated as

$$Q \approx \frac{(R-1) + L_1 F_2}{(R-1) + L'_1 F'_2} \tag{9.24}$$

Example 9.1

An RF engineer is considering installing an LNA at a CDMA base station to extend reverse-link range. The base station currently has a line loss of 2.0 dB between the antenna and the radio. The radio front end has a noise figure of 6.0 dB.

The LNA itself has a noise figure of 2.2 dB and a gain of 16 dB. The RF engineer would like to place the LNA on top of the tower right behind the antenna. This way, the loss between the antenna and the LNA would be 0.5 dB, and the loss between the LNA and the radio would be 1.5 dB. Assuming that the base station nominally experiences a reverse-link rise of 5.0 dB, what is the expected SNR improvement?

Solution

The following are given in the problem statement:

- $R = 5.0 \text{ dB} = 3.16$

- $L_1 = 2.0 \text{ dB} = 1.58$

- $F_2 = F'_4 = 6.0 \text{ dB} = 3.98$

- $F'_2 = 2.2 \text{ dB} = 1.66$

- $G'_2 = 16 \text{ dB} = 39.81$

- $L'_1 = 0.5 \text{ dB} = 1.12$

- $L'_3 = 1.5 \text{ dB} = 1.41$

Substituting these values into (9.23) yields an SNR improvement of 3.1 dB.

Since CDMA has reverse-link power control, *the improvement in SNR is manifested in the reduction of mobile transmit power.* This is intuitive because if there is an improvement in SNR, reverse power control would command the mobile to power down so that the mobile transmit power is just enough to achieve a desired link SNR.

Figure 9.4 shows the SNR improvement as a function of reverse-link rise and LNA noise figure. As expected, as reverse-link rise increases (i.e., more noise on the reverse link), the SNR improvement decreases. Also note that the SNR improvement increases as the LNA noise figure decreases.

Figure 9.5 shows the SNR improvement as a function of LNA gain and reverse-link rise. As expected, as reverse-link rise increases (i.e., more noise on the reverse link), the SNR improvement decreases. When the LNA gain increases, the SNR improvement also increases; however, the SNR improvement seems to reach saturation when the LNA gain reaches high levels.

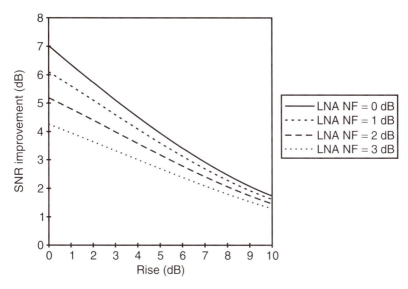

Figure 9.4 SNR improvement as a function of rise and LNA noise figure. The graph is generated using the following assumptions: $L_1 = 2.0\,dB$, $F_2 = F'_4 = 6.0\,dB$, $G'_2 = 16\,dB$, $L'_1 = 0.5\,dB$, and $L'_3 = 15\,dB$.

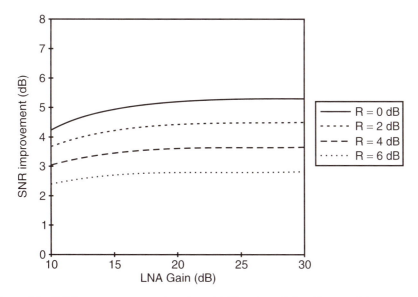

Figure 9.5 SNR improvement as a function of LNA gain and rise. The graph is generated using the following assumptions: $L_1 = 2.0\,dB$, $F_2 = F'_4 = 6.0\,dB$, $F'_2 = 2.2\,dB$, $L'_1 = 0.5\,dB$, $L'_3 = 1.5\,dB$.

9.4.4 Capacity Improvement

Field tests have shown that the mobile transmit power decreases when LNAs are installed at the serving base station. This phenomenon is due to reverse power control. LNAs decrease the noise figure of the receiving system and improve reverse link SNR. The manifestation of this improved link SNR is the decreased mobile transmit power. When reverse-link SNR improves, reverse-link power control directs the mobile to power down; this is done in order for the mobile to transmit just enough power to meet the required E_b/N_0 on the reverse link.

Although the mobile transmit power is decreased, the required E_b/N_0 is still the same on the reverse link. Recall from (4.14) in Chapter 4 that CDMA capacity is inversely proportional to the required E_b/N_0; that is

$$M \approx \frac{(W/R)}{(E_b/N_0)} \left(\frac{1}{1+\eta} \right) \lambda \left(\frac{1}{v} \right)$$

Since the required E_b/N_0 has not changed, the capacity of a base station with LNAs has not improved. However, the capacity of a neighboring base station (with no LNAs) may improve. Figures 9.6 and 9.7 illustrate the reason. In Figure 9.6, both base stations do not have LNAs. The mobile (served by base station 1) transmits at a nominal power level, and base station 2 is loaded by the transmit power of this mobile. In Figure 9.7, base station 1 now has LNAs installed, but base station 2 still does not have any LNAs. As a result, the transmit power of the mobile (served by base station 1) is lowered, and base station 2 consequently experiences a smaller loading by the mobile. In other words, the loading factor η has decreased for base station 2 as a result of installing LNAs at base station 1.

The net result is that the capacity of base station 2 is improved due to a smaller loading factor. This effect is not easily quantified, however, and the extent of the capacity improvement depends on the network configuration and the distribution of mobiles in the network.

9.5 Intermodulation

9.5.1 Intermodulation Theory

Intermodulation (IM) is a nonlinear process that generates an output signal containing frequency components not present in the input signal. Figure 9.8

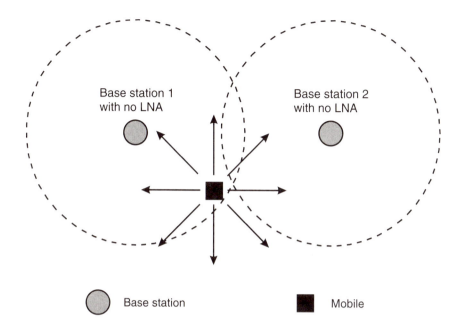

Figure 9.6 The mobile transmits at a nominal power level.

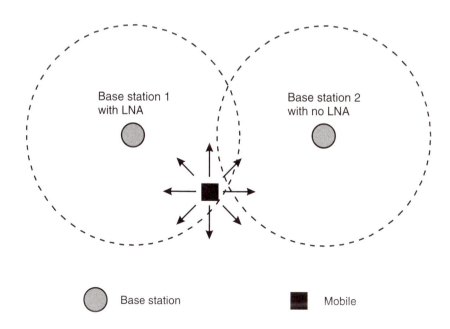

Figure 9.7 The mobile transmits at a lower power level.

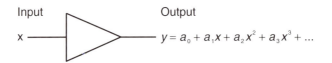

Input   Output

x ——————————▷—————— $y = a_0 + a_1x + a_2x^2 + a_3x^3 + \ldots$

Figure 9.8 A nonlinear device.

depicts the IM effect. The nonlinear device, in this case an amplifier, is characterized by the transfer function

$$y = a_0 + a_1 x + a_2 x^2 + a_3 x^3 + \ldots = y_0 + y_1 + y_2 + y_3 + \ldots \quad (9.25)$$

where x is the input of the amplifier and y is the output of the amplifier. If the amplifier were linear, then the transfer function would only contain the first two terms; that is,

$$y = a_0 + a_1 x = y_0 + y_1 \quad (9.26)$$

As we can see from (9.26), those frequency components present in x would also be present in y, and no new frequency components are generated. However, if the amplifier is nonlinear and contains higher order terms, then the output y would contain frequency components not originally present in x.

If we model the input x as two sinusoids at two different frequencies f_1 and f_2; that is,

$$x = A\cos(2\pi f_1 t) + B\cos(2\pi f_2 t)$$

then the output due to the second-order term (i.e., y_2) is

$$y_2 = a_2 x^2 = a_2 \left[A\cos(2\pi f_1 t) + B\cos(2\pi f_2 t) \right]^2$$

$$= a_2 \left[A^2 \cos^2(2\pi f_1 t) + 2AB\cos(2\pi f_1 t)\cos(2\pi f_2 t) \right.$$

$$\left. + B^2 \cos^2(2\pi f_2 t) \right] = a_2 \left\{ \frac{A^2}{2} \left[1 + \cos(2\pi f_1 t) \right] \right.$$

$$\left. + AB \left[\cos\left[2\pi(f_1 + f_2)t \right] + \cos\left[2\pi(f_1 - f_2)t \right] \right] \right.$$

$$+\frac{B^2}{2}\big[1+\cos(2\pi f_2 t)\big]\bigg\} = \frac{a_2}{2}\big(A^2+B^2\big)$$

$$+\frac{a_2}{2}\Big[A^2\cos\big(2\pi(2f_1)t\big)+B^2\cos\big(2\pi(2f_2)t\big)\Big]$$

$$+a_2 AB\Big\{\cos\big[2\pi(f_1+f_2)t\big]+\cos\big[2\pi(f_1-f_2)t\big]\Big\} \qquad (9.27)$$

The first term of (9.27) is a constant (DC term); the second term contains those frequency components that were not present in the input (i.e., $2f_1$ and $2f_2$). The third term also contains those frequency components that were not present in the input; these new frequency components are (f_1+f_2) and (f_1-f_2).

Generally, the output due to the second-order term does not cause too much of a problem. The reason is that the frequencies $2f_1$, $2f_2$, (f_1+f_2), and (f_1-f_2) are typically out-of-band and can be filtered out by bandpass filters. For example, if we approximate AMPS channels 414 and 354 on the forward link as two sinusoids with frequencies f_1 and f_2, then $f_1 = 882.42$ MHz and $f_2 = 880.62$ MHz. By evaluating the different frequency components due to the second-order term, we get

$$2f_1 = 2(882.42\text{ MHz}) = 1.76484\text{ GHz}$$

$$2f_2 = 2(880.62\text{ MHz}) = 1.76124\text{ GHz}$$

$$(f_1+f_2) = 1.76304\text{ GHz}$$

$$(f_1-f_2) = 1.80\text{ MHz}$$

As we can see, these out-of-band components can be easily filtered out by a bandpass filter (placed after the amplifier) that passes only the cellular band (see Figure 9.9).

The output due to the third-order term is

$$\begin{aligned}
y_3 = a_3 x^3 &= a_3\big[A\cos(2\pi f_1 t)+B\cos(2\pi f_2 t)\big]^3 \\
&= a_3\big[A^3\cos^3(2\pi f_1 t)+3BA^2\cos^2(2\pi f_1 t)\cos(2\pi f_2 t) \\
&\quad + 3AB^2\cos(2\pi f_1 t)\cos^2(2\pi f_2 t)+B^3\cos^3(2\pi f_2 t)\big]
\end{aligned}$$

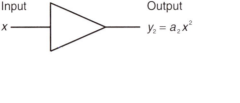

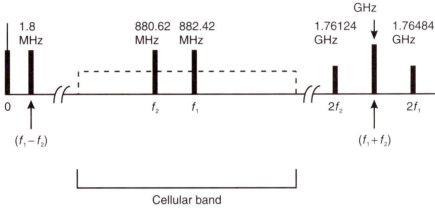

Figure 9.9 Second-order IM components. Note that $a_2 = 1$, $A = 1$, and $B = 1$.

$$= \frac{3a_3 A^3}{4} \cos(2\pi f_1 t) + \frac{a_3 A^3}{4} \cos(2\pi(3f_1)t)$$

$$+ \frac{3a_3 BA^2}{2} \cos(2\pi f_2 t) + \frac{3a_3 BA^2}{4} \cos(2\pi(2f_1 - f_2)t)$$

$$+ \frac{3a_3 BA^2}{4} \cos(2\pi(2f_1 + f_2)t) + \frac{3a_3 B^2 A}{2} \cos(2\pi f_1 t)$$

$$+ \frac{3a_3 B^2 A}{4} \cos(2\pi(2f_2 - f_1)t) + \frac{3a_3 B^2 A}{4} \cos(2\pi(2f_2 + f_1)t)$$

$$+ \frac{3a_3 B^3}{4} \cos(2\pi f_2 t) + \frac{a_3 B^3}{4} \cos(2\pi(3f_2)t) \tag{9.28}$$

By examining (9.28), we can see that there are six new frequency components that were not present in the original input. The frequencies and magnitudes of these components are

$$3f_1 \qquad \frac{a_3 A^3}{4}$$

$$3f_2 \qquad \frac{a_3 B^3}{4}$$

$$(2f_1 - f_2) \qquad \frac{3a_3 BA^2}{4}$$

$$(2f_1 + f_2) \qquad \frac{3a_3 BA^2}{4}$$

$$(2f_2 - f_1) \qquad \frac{3a_3 B^2 A}{4}$$

$$(2f_2 + f_1) \qquad \frac{3a_3 B^2 A}{4}$$

Generally, the frequency components $(3f_1)$, $(3f_2)$, $(2f_1 + f_2)$, and $(2f_2 + f_1)$ do not cause too much of a problem because they are out-of-band and can be filtered by a bandpass filter. However, the frequency components $(2f_1 - 2f_2)$ and $(2f_2 - f_1)$ can cause severe problem because these two components fall inside the operating band. Figure 9.10 shows the third-order IM components in the cellular band if $f_1 = 882.42$ MHz and $f_2 = 880.62$ MHz.

9.5.2 CDMA Scenario

In the case of cellular CDMA, intermodulation problems are prominent (on the forward link) when there are a lot of high-power AMPS carriers loading the CDMA phone. In cellular CDMA, the CDMA band is surrounded by AMPS carriers. As shown in Figure 9.11, if adjacent AMPS carriers are active, IM components could be generated by the front-end amplifier and fall within the CDMA band. These in-band IM components are interference within the CDMA band.

One way to reduce the amount of IM interference in the CDMA band is to minimize the power of AMPS carriers. If we reduce the input power (of AMPS carriers) by a factor of K, then the output due to the linear term is also reduced by a factor of K; that is,

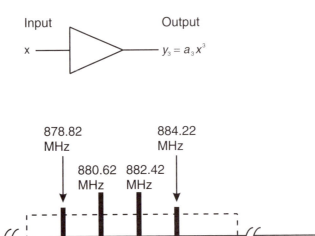

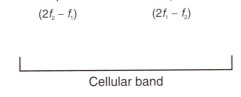

Figure 9.10 Third-order IM components in the cellular band. Note that $a_2 = 1$, $A = 1$, and $B = 1$.

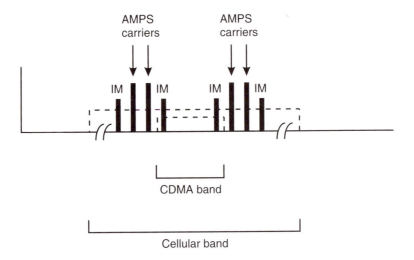

Figure 9.11 Third-order IM components are generated by adjacent AMPS carriers and fall within the CDMA band.

$$y_1 = a_1 \left(\frac{x}{K} \right) = \frac{a_1 x}{K}$$

However, the output due to the third-order term is reduced by a factor of K^3; that is,

$$y_3 = a_3 \left(\frac{x}{K} \right)^3 = \frac{a_3 x^3}{K^3}$$

To put it another way, if we attenuate the power of AMPS carriers by X dB, then the IM components would be attenuated by $3X$ dB. Thus, reducing the power of AMPS carriers drastically reduces IM interference. This can be accomplished by ensuring that AMPS channels do not operate at more than the necessary power level.

9.6 Interference Due to Other Mobiles

In Chapter 7, we described the reverse link E_b/N_0 in terms of (7.19):

$$\frac{E_b}{N_0} = \frac{T'L'_0(\theta_0, d_0)G_0(\theta_0)}{I'_m + I'_t + I'_n + N} \left(\frac{W}{R} \right)$$

The interference term I'_m is the total interference introduced by reverse traffic channel transmissions of all other mobiles in the same cell. This interference is a natural consequence of serving those mobiles in the same cell. The ability to minimize this interference is limited unless we can limit the number of mobiles that can be served by the home base station.

However, we can attempt to minimize I'_t. The interference term I'_t is the total interference introduced by reverse traffic channel transmissions of all other mobiles that are not served by the home base station. Reducing I'_t also implies reducing η in (4.14); that is,

$$M \approx \frac{(W/R)}{(E_b/N_0)} \left(\frac{1}{1+\eta} \right) \lambda \left(\frac{1}{v} \right)$$

Therefore, reducing I'_t means that we can increase the capacity of the current base station. We can accomplish this reduction by first examining the expression for I'_t. Equations (7.20) and (7.21) describe:

$$I'_t = \sum_{k=1}^{K} Y_k$$

where Y_k is the total traffic channel power (on the reverse link) received from those mobiles served by base station k. For base station k, Y_k is given by (7.21);

$$Y_k = \sum_{j=1}^{J_k} T'_{k,j} L'_{k,j} \left(\theta_{k,j}, d_{k,j} \right) G_0 \left(\theta_{k,j} \right)$$

The value $T'_{k,j}$ (reverse traffic channel ERP of mobile j) is controlled by reverse power control and limited by the maximum mobile transmit power. $L'_{k,j}(\theta_{k,j}, d_{k,j})$ is the reverse path loss. The antenna gain $G_0(\theta_{k,j})$ of the home base station is the only parameter that can be manually changed. By decreasing $G_0(\theta_{k,j})$, we can decrease Y_k and thus decrease the amount of interference I'_t received from those mobiles in neighboring cells. Downtilting an antenna is an effective way of reducing the receive gain to those mobiles that are far away while maintaining an adequate gain to those mobiles that are close by. In essence, downtilting decreases $G_0(\theta_{k,j})$ and reduces I'_t. Of course, by downtilting, one has to be careful not to adversely affect the receive footprint of the home cell.

References

[1]   Sklar, B., *Digital Communications: Fundamentals and Applications*, Englewood Cliffs, NJ: Prentice Hall, 1988.

Select Bibliography

Hamied, K., and G. Labedz, "AMPS Cell Transmitter Interference to CDMA Mobile Receiver," *Proc. 46th Annual Vehicular Technology Conf.*, IEEE, 1996, pp. 1467–1471.

Qualcomm, *The CDMA Network Engineering Handbook, Vol. 1: Concepts in CDMA*, 1993.

Simon, R. W., "Superconductor LNA Filters in Wireless Basestations," *Communication System Design*, July 1997, pp. 25-29.

SCT, "A Receiver Front End for Wireless Base Stations," Microwave J., April 1996, pp. 116-122.

TIA/EIA IS-95A, "Mobile Station-Base Station Compatibility Standard for Dual-Mode Wideband Spread Spectrum Cellular System," Telecommunications Industry Association.

TIA/EIA IS-98, "Recommended Minimum Performance Standards for Dual-Mode Wideband Spread Spectrum Cellular Mobile Station," Telecommunications Industry Association.

10

CDMA Traffic Engineering

10.1 Introduction

For cellular and PCS systems, traffic engineering is the process of provisioning communication circuits for a given service area. For a desired *grade of service*, the traffic engineer predicts the number of circuits necessary to satisfy the given traffic demand [1]. Communication circuits are typically provisioned at the base-station level. Although traffic engineering certainly applies to circuit provisioning beyond the base station (i.e., at the switching and interconnect levels), this chapter concentrates on traffic engineering at the base station level.

As an example, Figure 10.1 shows an AMPS base station that has an omnidirectional coverage pattern. The base station covers a major freeway and a shopping mall, and currently has 20 channels installed.

Given that there are 20 channels, it is intuitive to see that the traffic *demand* of this cell is directly related to the *blocking probability* of the cell. If there is more demand, then calls have higher probabilities of being blocked due to there being no channel available. If there is less demand, then calls have lower probabilities of being blocked.

Suppose that the base station currently experiences a very high *blocking rate*; then, in order to reduce the blocking rate, more channels must be installed at the base station.

This chapter presents the fundamental aspects of traffic engineering as related to a CDMA system. We first look at the traditional concepts of cellular traffic engineering. Then, we describe additional traffic engineering criteria that must be considered for a CDMA system.

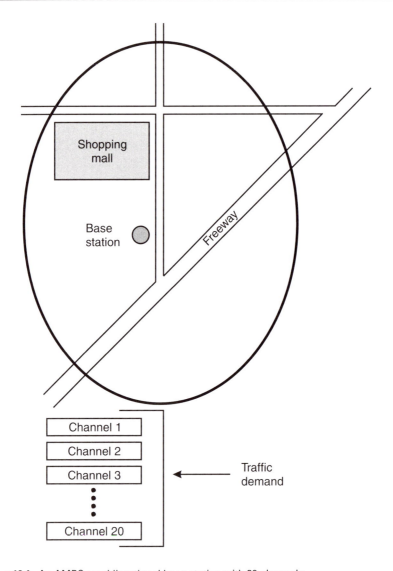

Figure 10.1 An AMPS omnidirectional base station with 20 channels.

10.2 Fundamental Concepts

10.2.1 Traffic Intensity

Before we start talking about traffic, we must first define the unit of measurement for traffic. Traffic is measured in terms of traffic *intensity*. Traffic intensity is commonly measured in the unit of *Erlang*. An Erlang is defined as the *average number of simultaneous calls.* For example, assume that the base station (shown in Figure 10.1) logs the number of active calls on a second-by-second basis; Table 10.1 shows such a log over a 10-sec interval.

In this case, the traffic intensity (the average number of simultaneous calls) is

$$\frac{(14 \text{ calls})(2 \sec) + (13 \text{ calls})(4 \sec) + (12 \text{ calls})(2 \sec) + (15 \text{ calls})(2 \sec)}{10 \sec}$$

$$= 13.4 \text{ Erlangs}$$

Note that the numerator is the total usage during this measurement interval (i.e., 10 sec), and that total usage during this measurement interval is 134 sec. Therefore, an equivalent definition of Erlang is the *total usage during a time interval divided by the time interval.*

Most network management systems do not measure the number of calls on a second-by-second basis. Doing so would require more processing time and memory. Thus, many network management systems measure usage during 1-hr intervals. For example, if the same base station logs a total usage of 72,000 sec (1,200 min) between 8:00 A.M. and 9:00 A.M., then the traffic intensity is

$$\frac{72,000 \sec}{3,600 \sec} = 20 \text{ Erlangs} \quad \text{or} \quad \frac{1,200 \min}{60 \min} = 20 \text{ Erlangs}$$

Because most network management systems measure usage during 1-hr intervals, an Erlang is sometimes referred to as 60 min or 1 hr of usage.

Table 10.1
A Hypothetical Log of the Number of Active Calls

8:00:01	14 calls
8:00:02	14 calls
8:00:03	13 calls
8:00:04	13 calls
8:00:05	13 calls
8:00:06	13 calls
8:00:07	12 calls
8:00:08	12 calls
8:00:09	15 calls
8:00:10	15 calls

10.2.2 Loads

In Section 10.1, we briefly referred to the traffic demand as some traffic requirement given by users in a cell. In reality, this demand cannot be directly measured, but only indirectly estimated. What we can directly measure is the *carried load* of a base station. For example, in the example shown in Section 10.2.1, the base station logged an actual traffic intensity of 20 Erlangs between 8:00 A.M. and 9:00 A.M. Thus, during this time interval, the base station has a carried load of 20 Erlangs. The traffic demand, or *offered load*, during the same time interval can be estimated by the following equation:

$$\text{carried load} = (\text{offered load}) \times (1 - \text{ blocking rate}) \qquad (10.1)$$

or

$$\text{offered load} = (\text{carried load}) / (1 - \text{ blocking rate}) \qquad (10.2)$$

The offered load is the amount of traffic load offered by users to the network, while the carried load is the amount of traffic load actually carried by the network. In other words, the carried load is the result of the offered load (demand) reduced by blocking. The network management system typically measures the rate of blocking for each base station in the system.

10.3 Grade of Service

We distinguish between the *blocking rate* and the *blocking probability*. Here, the blocking rate is defined as a measured quantity for a particular base station. On the other hand, the blocking probability is the probability that a call is blocked due to no channel available; this probability is a function of the desired offered load and the number of channels, and the probability can be evaluated using some mathematical model. As mentioned in Section 10.1, for a fixed number of channels, as the offered load (demand) increases, the blocking probability also increases. The term blocking probability is often used interchangeably with grade of service [2].

The blocking probability is typically evaluated for the offered load during the busy hour. For a base station, the busy hour is defined as the hour during which the highest carried load occurs. Figure 10.2 shows a typical carried load for a base station as a function of time. As we can see, the carried load on Thursday is higher than that on Monday. Because we want to ensure that the

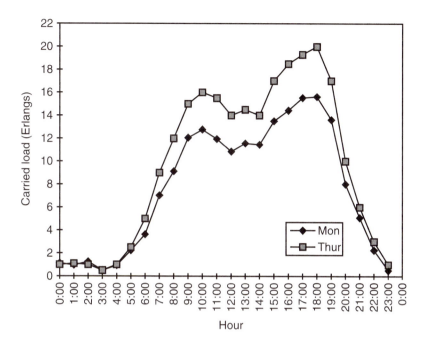

Figure 10.2 A typical intraday traffic distribution.

desired blocking rate is maintained during the busy time as well as during the nonbusy time, the channels are provisioned for the busy hour.

Erlang-B and Erlang-C are two widely used mathematical models that describe the relationship among the blocking probability (grade of service), offered load (demand), and number of channels. The models differ from each other in the underlying assumptions. We shall examine both Erlang-B and Erlang-C models.

10.3.1 Erlang-B Model

Erlang-B assumes that blocked calls are cleared and that the caller tries again later. In other words, the caller whose call is blocked does not immediately reoriginate the call. The blocking probability P(blocking), or grade of service, according to the Erlang-B model, is given by

$$P\left(\text{blocking}\right) = \frac{\dfrac{\rho^{C}}{C!}}{\displaystyle\sum_{i=0}^{C}\dfrac{\rho^{i}}{i!}} \tag{10.3}$$

where C is the number of channels and ρ is the offered load. Figure 10.3 depicts the relationship between the offered load and the number of channels for three different blocking probabilities. As we can see, the more channels there are, the more offered load a base station can handle for a desired blocking probability.

Some important assumptions used to derive (10.3) are as follows [2]:

- The system is in statistical equilibrium.
- The offered load is known.
- Calls arrive according to the Poisson process. This implies that the time between call arrivals is exponentially distributed. For this to be strictly true, a user whose call is blocked cannot immediately retry.
- Call service time is exponentially distributed.

Based on the last two assumptions, the offered load can be written as

$$\rho = \frac{\lambda}{\mu} \tag{10.4}$$

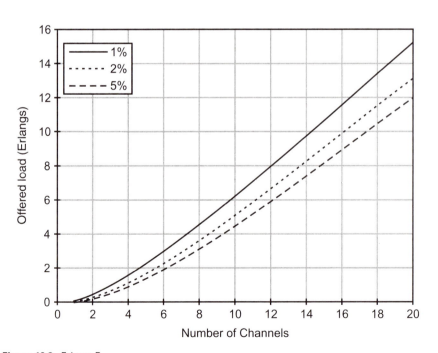

Figure 10.3 Erlang-B.

where λ is the Poisson arrival rate of λ calls/sec, and $(1/\mu)$ is the exponential call service time of $(1/\mu)$ seconds/call.

Example 10.1

Sector 1 of an AMPS base station has the carried load pattern as shown in Figure 10.2. The hour between 5:00 P.M. and 6:00 P.M. is the busy hour when the sector logged a carried load of 10 Erlangs. During the same hour, the blocking rate is logged at 9%. If the desired blocking rate is 1%, how many channels does the base station need (use Erlang-B)?

Solution

We first calculate the offered load:

$$
\begin{aligned}
\text{offered load} &= (\text{carried load}) / (1 - \text{blocking rate}) \\
&= (10 \, \text{Erlangs}) / (1 - 0.09) \\
&= 10.99 \, \text{Erlangs}
\end{aligned}
$$

By looking up the Erlang-B table or graph, we see that the base station needs 19 channels.

10.3.2 Erlang-C Model

At times, a user whose call is blocked continues to reoriginate until the call is established. Erlang-C assumes that blocked calls are retried until the call is established. Erlang-C models such retries as a queue (i.e., calls that are blocked are not lost, but rather are *delayed* until channels become available). The blocking probability according to Erlang-C is thus equivalent to the probability that the call is delayed, or the delay probability $P(\text{delayed})$. This probability is given by

$$
P(\text{delayed}) = \frac{\dfrac{\rho^C}{C!}}{\dfrac{\rho^C}{C!} + \left(1 - \dfrac{\rho}{C}\right) \displaystyle\sum_{i=0}^{C-1} \dfrac{\rho^i}{i!}} \tag{10.5}
$$

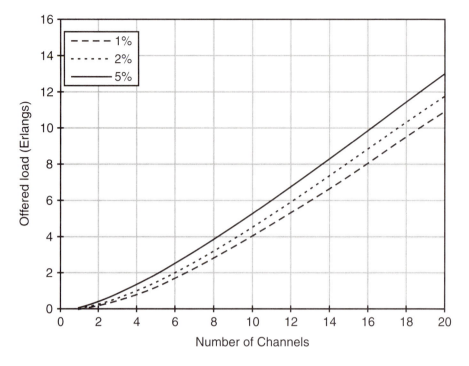

Figure 10.4 Erlang-C.

where C is the number of channels and ρ is the offered load. Figure 10.4 depicts the relationship between the offered load and number of channels for three different delay probabilities. As we can see from Figures 10.3 and 10.4, Erlang-C places a more stringent demand on circuit provisioning. Namely, Erlang-C requires more channels than Erlang-B to meet a given offered load and blocking probability.

Example 10.2

How many channels does the base station in Example 10.1 need if the desired blocking is still 1% (use Erlang-C)?

Solution

The offered load is still 10.99 Erlangs. By looking up the Erlang-C table or graph, we see that the base station needs 20 channels.

10.4 CDMA Applications

An analog base station blocks calls when there is no channel available. This form of blocking is called *hard blocking*. However, another blocking condition exists for a CDMA base station. Unlike AMPS and TDMA, CDMA does not impose a definite limit on blocking. As the number of users increases in a CDMA system, the level of interference increases as well, and this increase in interference negatively affects the quality of service. Because all users share the same RF spectrum, the interference increase contributes to a higher FER and a higher drop-call rate. In this case, the blocking is soft because the number of users can be increased if the service provider is willing to tolerate a higher level of interference and a lower quality of service. Soft blocking is a characteristic of the CDMA system. Therefore, two blocking scenarios exist for a CDMA base station:

- There may be plenty of channels available at a base station, but since there are many users in the same cell already, the interference level is such that adding an additional user would increase the interference above a predetermined threshold. The call is thus denied. This is the soft-blocking scenario.

- A call may have an excellent quality, but if there is no channel available at the base station, the call is still blocked. This is the hard-blocking scenario.

We describe each of the two blocking conditions in the following two sections.

10.4.1 Soft Blocking

We consider the soft blocking condition on the reverse link since this link is often the limiting link in terms of capacity [3]. We assume that there are a sufficient number of channels installed at the base station, so the probability of hard blocking is negligible.

We first describe a simplified model of soft blocking, and then we describe a more sophisticated model. Three assumptions are used in the simplified model; these assumptions are consistent with those used in Section 4.2:

1. There is a constant number of users M in the cell.

2. There is perfect power control.

3. Each user requires the same E_b/I_0.

Soft blocking occurs when the total interference level exceeds the background noise level by a predetermined amount $1/r$. First of all, we know that

$$\text{total interference} = (\text{same-cell interference}) + (\text{other-cell interference}) + (\text{thermal noise}) \tag{10.6}$$

In terms of CDMA parameters, (10.6) can be written as

$$I_{\text{total}} = M(E_b R) + \eta M(E_b R) + N \tag{10.7}$$

where

- M = number of users in the same cell;
- E_b = energy-per-bit of the signal;
- R = baseband data rate;
- N = thermal noise power;
- η = loading factor defined in (7.24) as the ratio of interference introduced by mobiles served by other cells to interference introduced by mobiles served by the home cell.

Equation (10.7) can be further manipulated into

$$I_{\text{total}} = M E_b R (1 + \eta) + N \tag{10.8}$$

As stated before, soft blocking occurs when the total interference level exceeds the background noise level by a predetermined amount $1/r$; thus, the condition for no soft blocking is

$$I_{\text{total}} \geq M E_b R (1 + \eta) + N \tag{10.9}$$

Given that

$$r = \frac{N}{I_{\text{total}}}$$

it follows that

$$M \leq \left(\frac{W/R}{E_b/I_0} \right) \left(\frac{1-r}{1+\eta} \right) \qquad (10.10)$$

Equation (10.10) describes the soft capacity M as a function of maximum allowable interference level.

In reality, none of the three previous assumptions holds because of the following:

1. The number of active calls is Poisson distributed with mean λ/μ.
2. Due to voice activity, each user is on with probability v and off with probability $(1-v)$.
3. Each user requires a different E_b/I_0 to achieve a desired FER.

We now use the above three conditions to generate a more sophisticated model of soft blocking. By using this new set of conditions, (10.9) can be written as

$$I_{\text{total}} \geq \sum_{i=1}^{m} \pi_i E_{b,i} R + \sum_{j=1}^{K-1} \sum_{i=1}^{m} \pi_{ij} E_{b,ij} R + N \qquad (10.11)$$

where m is the number of users in each sector and is assumed to be the same for all sectors, π_{ij} is the gating factor (due to voice activity) of the ith mobile in the jth sector, π_i is the gating factor of the ith mobile in the same (home) sector, and $E_{b,ij}$ is the energy per bit of the ith mobile in the jth sector. The first term on the right-hand side describes the interference introduced by those mobiles in the same cell, and the second term describes the interference introduced by those mobiles in other cells; the outer summation of the second term is over all other sectors (other than the home sector), and there is a total of K sectors in the system.

Dividing (10.11) by $I_0 R$ yields

$$(W/R)(1-r) \geq \sum_{i=1}^{m} \pi_i \left(\frac{E_{b,i}}{I_0} \right) + \sum_{j=1}^{K-1} \sum_{i=1}^{m} \pi_{ij} \left(\frac{E_{b,ij}}{I_0} \right) \qquad (10.12)$$

Note that m is a Poisson random variable with mean λ/μ, and π is a binary random variable with a value of 0 or 1; that is, π is the gating variable that has a value of 1 with probability v and a value of 0 with probability $(1-v)$

and thus v is effectively the voice activity factor. The value E_b/I_0 for each mobile depends on reverse-link power control and is random in nature; field results and simulation have shown that E_b/I_0 can be treated as a random variable with a log-normal distribution [3]. Since m, π, and E_b/I_0 are all random variables, the entire right-hand side can be treated as a random variable Z; that is,

$$Z \equiv \sum_{i=1}^{m} \pi_i \left(\frac{E_{b,i}}{I_0} \right) + \sum_{j=1}^{K-1} \sum_{i=1}^{m} \pi_{ij} \left(\frac{E_{b,ij}}{I_0} \right) \tag{10.13}$$

and the soft-blocking probability $P(\text{blocking})$ can be written as

$$P(\text{blocking}) = P\left[Z > (W/R)(1-r) \right] \tag{10.14}$$

Equation (10.13) is difficult to evaluate in closed form. Nevertheless, by making certain assumptions, one can approximate this expression in terms of (λ/μ). One such approximation is given by [3–5]

$$\frac{\lambda}{\mu} = \frac{(W/R)(1-r)}{(E_b/I_0)v(1+\eta)} F(B,\sigma) \tag{10.15}$$

where

$v =$ voice activity factor

$$F(B,\sigma) = \frac{1}{\alpha} \left[1 + \frac{\alpha^3 B}{2} \left(1 - \sqrt{1 + \frac{4}{\alpha^3 B}} \right) \right] \text{ in which}$$

$\alpha = \exp\left(\beta^2 \sigma^2 / 2 \right)$

$\beta = 0.2303$

$\sigma =$ standard deviation of power control

and

$$B = \frac{(E_b/I_0)\left[Q^{-1}(P(\text{blocking})) \right]^2}{(W/R)(1-r)}$$

Figure 10.5 shows the relationship between the maximum tolerable interference ratio ($1/r$) and soft capacity. The curve is calculated using a desired soft-blocking probability of 0.01 and the following assumptions:

- $W = 1.25$ MHz;
- $R = 9.6$ Kbps;
- $\sigma = 2.5$ dB;
- $\eta = 0.55$;
- $E_b/I_0 = 7$ dB;
- $v = 40\%$.

Example 10.3

We would like to assess the soft capacity of a CDMA base station. The system currently uses Rate Set 1 (9.6-Kbps full rate). The maximum tolerable interference ratio of total interference to background noise is 10. The standard deviation of reverse-link power control is 2.5 dB. The ratio of interference from mobiles served by other cells to interference from mobiles served by the home

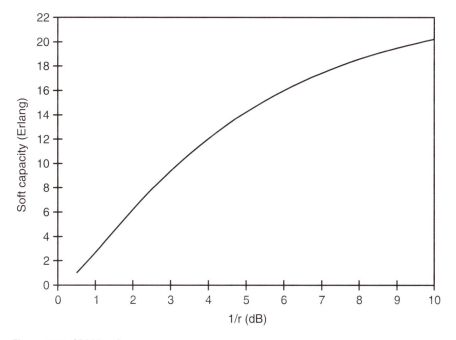

Figure 10.5 CDMA soft capacity.

cell is approximately 0.55. We would like to keep the soft-blocking probability at 1%.

We further make the assumption that E_b/I_0 is power-controlled to 7 dB most of the time, and the voice activity factor is 40%.

Solution

The following parameters are stated in the problem statement:

- $W = 1.25$ MHz;
- $R = 9.6$ Kbps;
- $W/R = 1.25 \times 10^6 / 9.6 \times 10^3 = 130.21$, which is the processing gain;
- $1/r = 10$; therefore, $r = 0.1$;
- $\sigma = 2.5$ dB or 1.778 as the standard deviation of reverse-link power control;
- $\eta = 0.55$;
- Desired $P(\text{blocking}) = 0.01$;
- $E_b/I_0 = 7$ dB or 5.012;
- $v = 40\%$ or 0.4.

We need to first assess $Q^{-1}(\text{blocking})$ or $Q^{-1}(0.01)$. By looking up a Q-function table, we obtain

$$Q^{-1}(0.01) = 2.33$$

Thus,

$$B = \frac{(E_b/I_0)\left[Q^{-1}(P(\text{blocking}))\right]^2}{(W/R)(1-r)} = \frac{(5.012)(2.33)^2}{(130.21)(1-0.1)} = 0.2322$$

$$\alpha = \exp(\beta^2\sigma^2/2) = \exp\left[(0.2322)^2(1.778)^2/2\right] = 1.0874$$

and

$$F(B,\sigma) = \frac{1}{\alpha}\left[1 + \frac{\alpha^3 B}{2}\left(1 - \sqrt{1 + \frac{4}{\alpha^3 B}}\right)\right]$$

$$= \frac{1}{1.0874}\left[1 + \frac{(1.0874)^3(0.2322)}{2}\left(1 - \sqrt{1 + \frac{4}{(1.0874)^3(0.2322)}}\right)\right]$$

$$= 0.5360$$

Now we can evaluate (λ/μ). It is

$$\frac{\lambda}{\mu} = \frac{(W/R)(1-r)}{(E_b/I_0)\nu(1+\eta)}F(B,\sigma) = \frac{(130.21)(1-0.1)}{(5.012)(0.4)(1+0.55)}(0.5360)$$

$$= 20.21\,\text{Erlangs}$$

10.4.2 Hard Blocking

We now consider the hard-blocking condition for a CDMA system. We assume here that the actual ratio of total interference to background noise is sufficiently small, so the probability of soft blocking is negligible. Consider a system consisting of three cells as shown in Figure 10.6. The coverage areas of the cells overlap and result in two-way soft handoff and three-way soft handoff areas.

Let x_1 be the percentage of time mobiles are not in soft handoff, x_2 be the percentage of time mobiles spend in two-way soft handoff, and x_3 be the percentage of time mobiles spend in three-way soft handoff. Hence

- x_1 = percentage of time mobiles are not in soft handoff;
- x_2 = percentage of time mobiles spend in two-way soft handoff;
- x_3 = percentage of time mobiles spend in three-way soft handoff;
- $x_1 + x_2 + x_3 = 100\% = 1.00$

Thus, x_1 can be written as

$$x_1 = 1 - x_2 - x_3$$

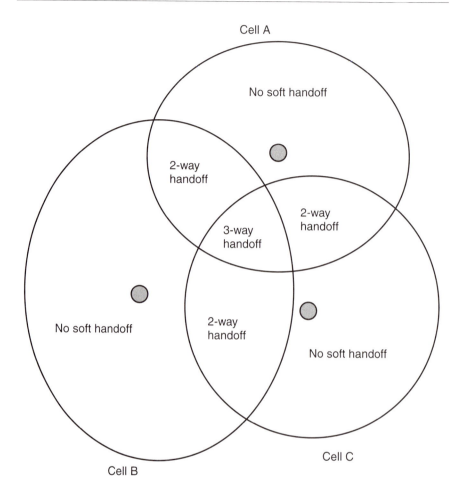

Figure 10.6 Three overlapping CDMA cells with areas of two-way soft handoff and three-way soft handoff.

Here we distinguish between *apparent* load and *real* load. In AMPS, apparent load and real load are equivalent. In CDMA, apparent load is the amount of load that is billed to the users, while real load is the amount of actual load experienced by the cellular or PCS system. The difference between these two types of loads is due to soft handoff. During a two-way soft handoff, the mobile is utilizing channel resources from two different cells at the same time, and during a three-way soft handoff, the mobile is utilizing channel resources from three different cells at the same time. Therefore, the *load factor*, defined as the ratio of real load to apparent load, is given by

$$\text{load factor} = \frac{\text{real load}}{\text{apparent load}}$$

$$= \frac{x_1 + 2x_2 + 3x_3}{x_1 + x_2 + x_3} = \frac{(1 - x_2 - x_3) + 2x_2 + 3x_3}{1}$$

$$\text{load factor} = 1 + x_2 + 2x_3 \qquad (10.16)$$

and

$$\text{real load} = \text{load factor} \times (\text{apparent load})$$

Therefore, real load should always be used to provision the number of channels in a CDMA system.

Example 10.4

Through market research and demographic study, the traffic engineer estimates that the offered load for a new cell is 11 Erlangs. The engineer further estimates that the percentage of time mobiles spend in two-way soft handoff is 30% and the percentage of time mobiles spend in three-way soft handoff is 10%. If 1% is the desired hard-blocking probability, how many traffic channels should be provisioned at the base station (using Erlang-C)?

Solution

The load factor is

$$\text{load factor} = 1 + 0.3 + 0.1 = 1.4$$

The real offered load is thus

$$\text{real offered load} = 1.4(11 \text{ Erlangs}) = 15.4 \text{ Erlangs}$$

Looking up 15.4 Erlangs in the Erlang-C table at 1% blocking, we obtain 26 traffic channels.

References

[1]   Faruque, S., *Cellular Mobile Systems Engineering*, Norwood, MA: Artech House, 1996.

[2]   Hess, G. C., *Land-Mobile Radio System Engineering*, Norwood, MA: Artech House, Inc., 1993.

[3]   Viterbi, A. M., and A. J. Viterbi, "Erlang Capacity of a Power Controlled CDMA System," *IEEE J. on Selected Areas in Communications*, Vol. 11, No. 6, Aug. 1993.

[4]   Viterbi, A. J., *CDMA Principles of Spread Spectrum Communication*, Reading, MA: Addison-Wesley, 1995.

[5]   Garg, V. K., K. Smolik, and J. E. Wilkes, *Applications of CDMA in Wireless/Personal Communications*, Upper Saddle River, NJ: Prentice Hall, 1997.

Select Bibliography

Everitt, D., "Traffic Engineering of the Radio Interface for Cellular Mobile Networks," *Proc. IEEE*, Vol. 82, No. 9, Sept. 1994, pp. 1371–1382.

Lee, W.C.Y., *Mobile Communications Design Fundamentals*, New York, NY: John Wiley & Sons, 1993.

Leung, K. K., W. A. Massey, and W. Whitt, "Traffic Models for Wireless Communication Networks," *IEEE J Selected Areas of Communication*, Vol. 12, No. 8, Oct. 1994, pp. 1353–1364.

Mehrotra, A., *Cellular Radio, Analog and Digital Systems*, Norwood, MA: Artech House, 1994.

11

Management Information Systems for Personal Communication Networks

11.1   Introduction

A cellular or PCS network does not exist in a vacuum. A network is installed and deployed in order to provide on-demand wireless communication services to individual users. As such, the network needs to be continuously maintained. The operation and management of a network require that a service provider know the state of its system in a timely manner.

A variety of information on the status of the system is necessary to operate and manage a personal communication network (see Figure 11.1). The ultimate goal of maintaining the network is to provide end users with the best mobile communication service. For example, if a particular cell in the network stops taking traffic, the event impacts those users in the vicinity of that cell since they would not be able to originate calls. Some information about the nature of the problem would influence the action taken by the service provider. If there is a power outage, then a technician would be dispatched to repair the power plant of that cell. If the communication link between the cell and mobile switching center is down, then the operator of that communication link (e.g., the local telephone company) would be contacted to repair the link. Obtaining timely information on the state of the network enables a service provider to best operate and manage its network assets.

Nevertheless, a personal communication network is a complex system. A variety of information on its configuration and performance needs to be constantly collected, processed, and stored. This chapter addresses some of these information needs for the operation and management of a personal

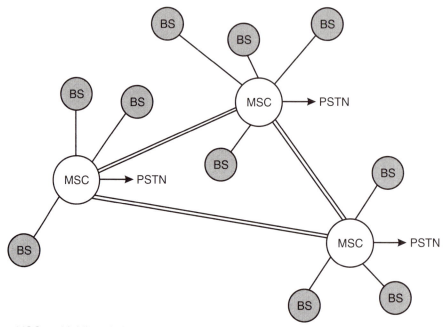

MSC: Mobile switching center
BS: Base station
PSTN: Public switched telephone network

Figure 11.1 A cellular or personal communication network.

communication network. We formalize these information requirements from the perspective of *management information systems* (MIS). We first review a model of management information systems; then, we apply this model to various operation and management information requirements of a cellular or personal communication system.

11.2   Management Information Systems

11.2.1 Information System and Control

Management information systems refer to computer-based information processing systems supporting the operation, management, and decision functions of an organization. Information systems in organizations, such as network engineering of a wireless service provider, provide information support for decision-makers at various management and decision levels. We define *information* as data that has been processed and is meaningful to a user [1]. In this chapter, we focus only on the information requirement of the network engineering function of a wireless service provider.

As shown in Figure 11.2, data on the state of the network is continuously collected from the physical *network* by the *management information system*. The raw state data is stored in the *database* and processed by the *model base* of the information system. We distinguish between the database and the model base of an information system; the database is the collection of all data items, while the model base is a collection of application programs that utilize the database [1]. The information necessary for the smooth operation and management of a network is the output of the management information system. The decision-maker utilizes the information in order to determine whether or not some corrective action should be taken.

For example, a common type of raw data produced by the network and collected by the management information system is the count of the number of blocked calls for a particular cell. Another common type of raw data is the count of the number of access attempts received by a particular cell. These two counts are collected by the management information system and stored in its

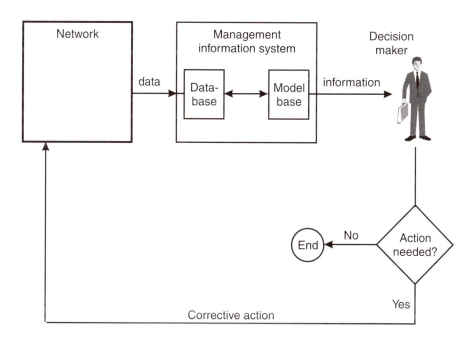

Figure 11.2 The role of the management information system in the maintenance of a personal communication network.

database. The model base contains an application used to generate network performance reports. The application retrieves these two counts over the last 24 hours; it divides the blocked-call count by the access-attempt count to arrive at a *blocking rate* for that particular cell. This information is examined by the decision-maker, typically an RF engineer. If the blocking rate is sufficiently low, the engineer may decide to take no action. However, if the blocking rate is too high (e.g., more than an objective of 2%), the engineer may decide to install additional radio channels at that particular cell. The installation of additional channels is the corrective action taken by the decision-maker.

The process shown thus constitutes a feedback control system. The essence of control includes sensing the system's outputs, comparing them to the objectives, and generating a corrective action, if necessary. Therefore, the network is a system that is being *controlled*, while both the management information system and the decision-maker constitute the *control* system.

The corrective action taken usually takes the form of a change in the inputs to the controlled system, but it may also lead to a change in the structure of the controlled system or to changed objectives [1]. For example, the objective of a service provider is to keep the blocking rate below 2%. If adding channels cannot achieve this goal, then the service provider may decide to add an additional cell in the vicinity to alleviate traffic congestion (i.e., a change in the structure of the controlled system). However, if the service provider decides that the 2% objective cannot be achieved without making excessive investments in the infrastructure, then it may decide to relax the 2% objective to 3% (i.e., changed objectives).

11.2.2 Classes of Decisions

There are three different classes or types of decisions made by decision-makers. These different decision classes are differentiated based on whether or not the decision-making process is *algorithmic* or *heuristic*. These three different decision types are as follows:

1. A *structured decision* is one where all the decision-making steps are structured. As such, the entire decision-making process can be formalized in an algorithm and programmed into a computer.

2. An *unstructured decision* is one where all the decision-making steps are unstructured. Thus, the decision-making process is heuristic and cannot be formalized in an algorithm.

3. A *semistructured decision* is one where some decision-making steps are structured and some are unstructured. As a result, some steps in

making a semistructured decision can be algorithmic and programmed, while some steps cannot be programmed [1–3]. Strictly
speaking, *decision support systems* (DSS) are a class of systems that
support semistructured decisions.

The decision to add additional channels described in the previous section
is an example of a semistructured decision. It is so because part of the decision-
making process is algorithmic; that is, if the blocked-call count divided by the
access-attempt count exceeds 2%, then the decision-maker decides whether or
not to take a corrective action.

However, there are also instances where structured decisions would be
made on the network. For example, if the power amplifier of a cell exceeds its
rated wattage, then an overload condition exists and a decision should be made
to shut down the amplifier. This decision to take a corrective action is algorithmic and requires no human intervention. Thus, an application can be written for the model base of a management information system to automatically
shut down the power amplifier if there is an overload condition. The decision-
making process shown in Figure 11.2 is thus modified to include the structured
decision-making process. The modified process is shown in Figure 11.3.

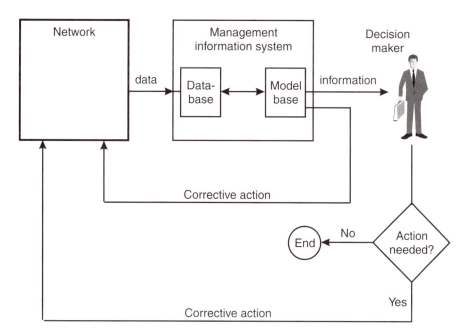

Figure 11.3 The modified role of the management information system in the maintenance
of a personal communication network.

Most of the operation and management decisions made in the network engineering function are semistructured and structured. These decisions almost always require information that is produced by the management information system. Unstructured decisions regarding the network, such as strategic decisions on future network evolution, are not addressed in this chapter.

11.3   Network Management

As mentioned before, the engineering decisions made regarding the network almost always make use of information produced on the state of the network. The operation and management of the network can be divided into five different areas:

- Fault management;
- Performance management;
- Configuration management;
- Planning;
- Call accounting.

Different databases and model bases exist in the management information system to support these operation and management functions.

11.3.1 Fault Management

As described before, an element of a personal communication network may fail at any time, and the failure of such an element may have major impacts on customers. Thus, fault management is necessary to monitor the status of the network and to alert the service provider of any failures or faults that have occurred in the network. Fault management typically consists of the following functions as shown in Figure 11.4.

The fault detection function monitors the alarms that occur in the network, identifies which network element generated a given alarm, and ranks each occurring alarm in order of priority. For example, if the communication link between a cell and its mobile switching center fails, this failure is immediately customer-impacting and thus should be given a higher priority; if a single channel card goes out of service, this failure is not immediately customer-impacting and thus should be given a lower priority.

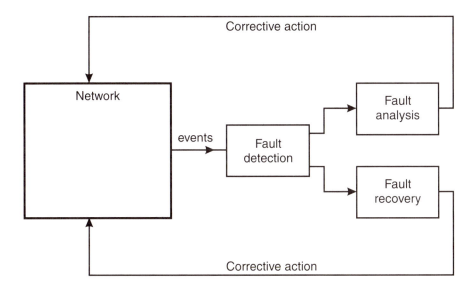

Figure 11.4 Fault management.

The fault recovery function involves those processes necessary to correct a fault or recover from a failure. The corrective actions taken by this function may be automatic, manual, or a combination of both. For example, if the primary tower-top amplifier of a cell fails, then the redundant tower-top amplifier should be automatically switched into the receive path. This decision is structured and automatic, and requires no human intervention. On the other hand, if a cell goes out of service due to a power outage, then a technician is dispatched to repair the power plant; this is a manual process.

The following is an example involving both automatic and human interventions. If a channel card fails, then the fault recovery function should automatically take it out of service; at the same time, a technician should be dispatched to swap out the bad channel card.

The fault analysis function analyzes the records of all fault occurrences in the network and produces summary reports on the fault history. This function enables the service provider to identify any short-term or long-term trends in fault occurrences. For example, if a CDMA cell consistently experiences "reverse-link noise" alarms (triggered if the total receive power goes above a predetermined threshold), then there may exist a (intentional or unintentional) jamming transmitter in the vicinity of the cell. In this case, a noise surveillance team should be sent to the area to identify and locate the jammer.

11.3.2 Performance Management

Performance management refers to the continuous collection and processing of traffic and quality-of-service information on the network. Performance management contains three main functions; they are performance monitoring, performance optimization, and trend analysis (see Figure 11.5).

The performance monitoring function includes processes that continuously collect traffic and quality-of-service information on the network. For example, on a per-sector basis, the performance monitoring function measures the traffic usage in Erlangs for each sector (i.e., traffic information). It also measures the blocking and drop-call rates for each sector (i.e., quality-of-service information).

The performance optimization here almost always involves semistructured decision-making. By monitoring the performance information, the decision-maker (i.e., the engineer) ascertains whether or not a particular performance metric meets the organizational objective. For example, if the blocking rate of a cell is 7% while the organizational objective is 2%, then the engineer may decide to install additional channels at that cell in order to decrease blocking. For an analog system, the corrective action would involve frequency planning and retuning in order to add additional channels. For a CDMA system, further analysis is carried out to determine whether or not a soft- or hard-blocking condition exists (see Chapter 10).

The performance optimization function within performance management also has the ability to make changes to specific network parameters on-

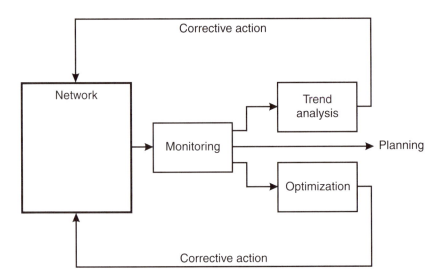

Figure 11.5 Performance management.

line. For example, if the performance monitoring function shows that the soft-handoff statistics for a cell are too high, then the engineer may decide to decrease the number of soft handoffs occurring on this particular cell. In doing so, he or she decreases the pilot power parameter or increases T_ADD and T_DROP threshold parameters of that cell. These parameter changes (i.e., corrective actions) can typically be made on-line as part of the performance optimization function.

The trend analysis function examines statistics collected by the performance monitoring function over a longer period of time to determine whether or not an underlying performance trend exists. A corrective action can also be taken as a result of trend analysis. For example, say after the engineer installs additional channels, the blocking rate of the previously mentioned cell still remains above 2% over a one-month period. This long-term blocking trend may indicate a generally high traffic demand in the area; thus, as a result of the trend analysis, the engineer decides to add an additional cell in the vicinity to meet the high traffic demand.

11.3.3 Configuration Management

Garg [4] defines configuration management as a set of functions and processes used to identify and control the different network elements. These functions include, but are not limited to, the following:

- A procurement function that purchases different hardware and software necessary for the smooth operation and management of the network. For example, a process has to exist to purchase base station equipment, antennas, and software diagnostic tools.

- An inventory function that tracks the different network hardware and software elements. For example, a service provider should know exactly how many channel cards there are in the network. These channel cards may include those that are installed and in service, installed and not in-service, deinstalled and in working condition, and deinstalled and waiting to be repaired. Knowing exactly what and where the usable network resources are enables a service provider to efficiently maintain its network.

- A tracking function that keeps a log of different changes made to the network. For example, if an engineer requests to change the antennas of a particular cell, a tracking process should exist to record when the request was made, the nature of the request (i.e., which antennas are to be replaced and what antennas will replace them), and when the change work was carried out and completed. The tracking function permits a service provider to effectively control its network and allocate resources within its network.

11.3.4 Planning

The planning function deals with the consideration of the future configuration of the network in response to market demands. The decisions relating to this function include how many additional cells need to be built in the next five years and how many mobile switching centers are needed to support that many cells in the next five years. Planning typically considers the configuration of the network two to five years in the future.

The planning function takes as its input traffic information from performance management and configuration information from configuration management. Planning may also consider other inputs, such as marketing goals (i.e., how many additional subscribers are going to be captured in the next five years). As Figure 11.6 shows, these inputs are fed into the forecast function. This function analyzes historical traffic information and takes into account future marketing goals; then, it forecasts the total network traffic in the future years. By using the current network configuration and traffic distribution among the current cells, the forecast outputs the projected traffic load for each cell (sector) in the network.

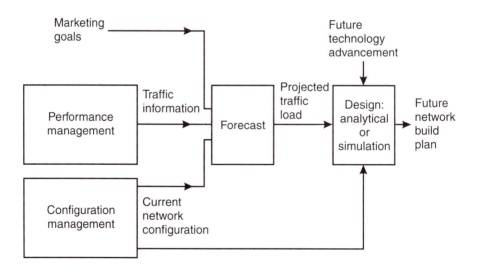

Figure 11.6 System planning.

These projected traffic loads serve as inputs to the design function. This function involves a semistructured decision-making process that determines how many additional cells are to be built in the future and where the additional cells will be. For an analog system, the process is straightforward; it typically involves examining the hard capacity of each cell and comparing the hard capacity with the projected traffic load for that cell. If the projected traffic load is greater than the hard capacity of that cell, then an additional cell needs to be built in the vicinity.

For a CDMA system, the process is more complicated because it involves determining the soft capacity of the cell. *Analytical* methods such as those described in Chapter 10 exist to calculate the soft capacity. CDMA *simulation* can also be used to assess the soft capacity of a cell and identify the capacity-limiting areas in the network. By using analytical methods, we can ascertain the soft capacity of the cell, and if the projected traffic load is greater than the soft capacity, then we know that an additional cell needs to be built in the vicinity. By using simulation, we can identify the capacity-limiting areas in the system, and additional cells typically need to be built near these capacity-limiting areas. Note that the design function also takes into account possible future advancements in technology and their effects on the eventual network configuration. For example, adaptive beam-forming (i.e., smart antenna) for CDMA will increase the capacity per cell; thus, if this technology becomes available in the future, fewer cells (with adaptive beam-forming) will be needed to support a given traffic demand.

The output of the planning function is a set of network build plans for the future years.

11.3.5 Call Accounting

Call accounting refers to those portions of the management information system that produce usage and billing information for individual customers. It also contains applications that compile demographics and customer profile information, as well as applications that detect fraudulent usage patterns and alert the service provider that a particular account may have been cloned. Call accounting includes four different functions, as described below [4].

The metering function creates the usage-metering records that are generated every time a call is originated or received, or every time a billable service is requested, such as the 411 information service.

The charging function compiles the usage-metering records generated by different sources and creates a *call detail record* for each call that occurs on the network. The call detail record contains accounting (e.g., duration) as well as technical details (e.g., originating and terminating cells) regarding each call.

The billing function further processes the call detail records into individual bills for individual customers. It does so by collecting call detail records belonging to a particular customer during a billing period and generating bills for that particular customer.

The analysis function processes groups of call detail records in order to produce analysis reports. These reports may contain usage pattern reports indicating who the high-usage customers are. This function also analyzes the usage pattern and estimates which accounts most likely contain fraudulent usage. Figure 11.7 depicts the call accounting function.

11.4 Concluding Remarks

Needless to say, the different functions of a management information system necessary to maintain a network are complex. The amount of information existing on the state of the network is enormous, and careful planning and design of the management information system are essential to meeting operation and management requirements and accommodating future growth of the network.

Although the management information system provides much of the needed information on the network, we should mention that the management information system alone is not sufficient to operate and manage a large and complex network. For example, if the trend analysis function of performance management indicates that a cell consistently has high drop-call rates, this

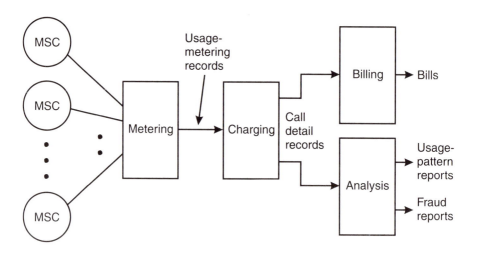

Figure 11.7 Call accounting.

information indicates that there is a network problem but does not reveal the cause of the problem. There may be a variety of causes: the cell may have reached its capacity, the pilot power may be too high and cause mobiles to originate from far away, or the antenna connection may simply be bad. In this case, additional information is needed to decide which corrective action to take. The additional information may be obtained by test-driving the system or by sending a technician out to the cell to check antenna connections.

In this regard, the network information produced by the management information system is equivalent to a *resource*. Although this resource is vital to the operation and management of the network, other resources, such as humans and diagnostic equipment, are also necessary to maintain the network.

References

[1]   Ahituv, N., S. Neumann, and H. N. Riley, *Principles of Information Systems for Management*, Dubuque, IA: Wm. C. Brown Communications, Inc., 1994.

[2]   Simon, H. A., *The New Science of Management Decisions*, New York, NY: Harper & Row, 1960.

[3]   Neumann, S., and M. Hadass, "DSS and Strategic Decisions," *California Management Review*, Vol. 22, No. 2, Spring 1980, pp. 77–84.

[4]   Garg, V. K., K. Smolik, and J. E. Wilkes, *Applications of CDMA in Wireless/Personal Communications*, Upper Saddle River, NJ: Prentice Hall, 1997.

Select Bibliography

CTIA, "Requirements for Wireless Network OAM&P Standards," CTIA OAM&P SG/95.11.28.

ITU Recommendation X.721, "Definition of Management Information," 1992.

ITU Recommendation M.3100, "Generic Network Information Model," 1992.

ITU-T Recommendation M.3010, "Principles for a Telecommunications Management Network," 1996.

Terplan, K. *Communication Networks Management*, Upper Saddle River, NJ: Prentice Hall, 1992.

12

RF Regulatory Considerations

12.1 Motivation

The *Federal Communications Commission* (FCC) regulates the use of the RF spectrum in the United States. It is the sole government body responsible for the regulation of these frequencies, including frequency assignment, operation guidelines, and rules enforcement. In engineering a cellular or PCS system, the regulatory guidelines set by the FCC must be carefully followed, since the FCC has not only the authority to grant cellular and PCS operating licenses but also the power to revoke them. In extreme cases, criminal prosecution may be pursued for violating statutes.

One of the most important rules to follow is making sure that the *service area boundary* (SAB) of a mobile system does not overstep the carrier's *cellular geographic service area* (CGSA), assigned as part of the operating license. The CGSA of a cellular carrier is the area considered by the FCC to be served by the carrier's mobile system. The CGSA is also the area within which the carrier is entitled to protection from adverse effects, one of which may be interference from neighboring carriers operating in the same frequency block. This chapter first of all discusses various issues involved in determining the SAB of a CDMA cellular system. We first review the conventional technique of calculating SAB boundaries for an AMPS system. Because CDMA is intrinsically a different technology from AMPS, the field strength method of determining service boundary does not translate directly into a digital modulation scheme whose information content is spread across a much wider bandwidth. We then develop several alternative methods of calculating the SAB for a CDMA system based on the meaningful definition of service area boundary.

Second of all, another important rule to follow is the *maximum permissible exposure* (MPE) limitation. The FCC has set forth *electromagnetic energy* (EME) guidelines that must be carefully followed. This chapter also reviews the general guidelines of engineering a cell site to comply with these rules.

12.2 SAB Determination

The FCC considers cellular service to be provided in all areas between the base station and the locus of points where the predicted or measured median field strength decreases to 32 dBμV/m (or dBμ) [1]. However, the derivation of this field strength value is based on the conventional analog FM scheme whose bandwidth is 30 kHz. For a digital CDMA system whose carrier bandwidth is effectively 1.25 MHz, an alternative metric is needed to meaningfully define the SAB. In this section, we first review the method of SAB calculation for an AMPS system; then, we consider three alternative methods of calculating the SAB for a CDMA system. Note that these methods are technical alternatives and have not been endorsed by the FCC. Each service provider is responsible for using appropriate methods to determine its SAB.

12.2.1 Review of AMPS SAB Calculation

The method used to derive SAB field strength for an FM system was first popularized by Carey [2]. One of the fundamental assumptions in [2] is that "only the reception of a base station by a mobile unit is considered, since this is the controlling factor for assignment purposes." While this forward-link assumption may be adequate in the AMPS case, a CDMA system is either forward- or reverse-link limited (depending upon the vocoder rate), and this link limitation should be taken into account in determining the SAB for a CDMA system.

The actual value of 32 dBμ for minimal required field strength was later proposed by the FCC and formalized by Lee [3] and others. In the conventional AMPS system, normal cellular practice is to specify a *C/I* of 18 dB for acceptable voice-quality level [4]. By defining a commonly accepted noise floor level as a standard quantity, the carrier power required to attain a *C/I* of 18 dB can be calculated, and this required carrier power can then be converted to a field strength quantity. The noise power *N* present in a channel bandwidth of *W* is

$$N = kTW = N_0 W \tag{12.1}$$

where k is Boltzmann's Constant (-228.6 dBW/Hz/K). The quantity N equals -129 dBm for a channel bandwidth of 30 kHz at 290K. However, field measurements have shown that actual noise level is typically 8 to 11 dB higher than the theoretical noise level due to other man-made noises [3]. Therefore, if we use a noise rise level of 9 dB, the total interference level $I = -129$ dBm + 9 dB = -120 dBm. Given a C/I of 18 dB, the required carrier power is then $C = 18$ dB $- 120$ dBm = -102 dBm (or 6.31×10^{-14} W).

To convert the required carrier power of -102 dBm into field strength, we invoke Poynting's theorem:

$$\mathbf{P} = \mathbf{E} \times \mathbf{H} \qquad (12.2)$$

where $\mathbf{P}$ is the Poynting vector that represents the rate of energy flow (power) out through a surface. In free space, $\mathbf{E}$ is perpendicular to $\mathbf{H}$, and

$$E = \eta_o H \qquad (12.3)$$

where $\eta_o = 376.73\Omega$ and is known as the intrinsic impedance of free space. Substituting (12.3) into (12.2) yields

$$P = \frac{E^2}{\eta_o} \qquad (12.4)$$

or

$$E = \sqrt{\eta_o P} \qquad (12.5)$$

Equation (12.5) expresses the electric field E as a function of power flow per unit area P. If we assume a half-wave dipole antenna, then the power p_r intercepted (received) by the dipole antenna with an effective aperture area A is given by

$$p_r = PA \qquad (12.6)$$

The aperture area A can be derived from the universal antenna formula:

$$G = \frac{4\pi A}{\lambda^2} \qquad (12.7)$$

With a half-wave dipole, the gain G is 1.64 times (or 2.15 dB) above isotropic; therefore,

$$A = \frac{1.64\,\lambda^2}{4\pi} \tag{12.8}$$

Combining (12.8), (12.6), and (12.5) yields

$$E = \sqrt{\frac{4\pi\eta_o p_r}{1.64\,\lambda^2}} = \sqrt{\frac{4\pi\eta_o p_r f^2}{1.64\,c^2}} \cong 1.79 \times 10^{-7} f \sqrt{p_r} \tag{12.9}$$

For example, for a minimum required carrier power of 6.31×10^{-14}W (-102 dBm) and a frequency of 881.5 MHz, the corresponding electric field E is 39.7 μV/m, or 32 dBμ (see Figure 12.1).

The distance from a periphery base station to where the field strength decreases to 32 dBμ, thus by definition to its SAB, is partly predicted by the

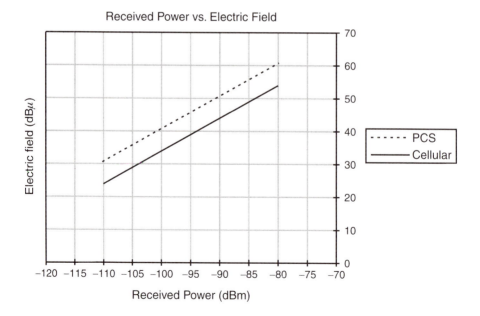

Figure 12.1 The derived relationship between received power and electric field strength for signals at 881.5 MHz (cellular) and 1.920 GHz (PCS).

medium field strength curves in [2]. The FCC has adopted the use of a parameterized equation, predicting such a distance as a function of base station ERP and antenna radiation center *height above average terrain* (HAAT):

$$d = 2.531h^{0.34}p^{0.17} \tag{12.10}$$

where d is the radial distance in kilometers, h is the antenna HAAT in meters, and p is the ERP in watts.

12.2.2 CDMA SAB Determination With Multiple Sectors

Unlike its AMPS counterpart, the CDMA system as specified in IS-95 does not use the same modulation schemes on both the forward and reverse links. The forward link consists of four channel types—pilot, paging, sync, and traffic channels—and the reverse link consists of two channel types—access channels and traffic channels. Of the four forward channel types, the pilot channel and its signal effectively determine the forward coverage area of a base station. The pilot signal is transmitted by a base station at a relatively higher level than any other channel. A call cannot be set up without the mobile's receipt of sufficient pilot strength because, along with other functions, the pilot is used as a coherent carrier phase reference for demodulation of the other channel signals from the serving base station.

Although the pilot signal determines the forward coverage area, it does not alone define the effective service area of a base station. In addition to its successful demodulation of the pilot signal, the mobile must transmit enough power to close the reverse-link traffic channel. And in some situations, the CDMA system is reverse-link limited. Therefore, a meaningful definition of SAB for a CDMA system should take into account both the *forward* pilot channel and the *reverse* traffic channel. We define d_f, or forward service distance, to be the radial distance between a base station and a mobile such that the received pilot strength is above a predetermined threshold, and d_r, or reverse service distance, to be the radial distance such that the received traffic signal strength is above a predetermined threshold. Thus, for J radials, the forward SAB for a base station (or sector) in a multiple cell environment is defined as the locus of points q_1 to q_J such that $q_j = d_{f,j}$, and the reverse SAB is defined as the locus of points q_1 to q_J such that $q_j = d_{r,j}$.

The forward pilot strength is defined in terms of a ratio of pilot chip energy to total received power spectral density (E_c/I_0); this quantity for a probe mobile is

$$\frac{E_c}{I_0} = \frac{\alpha_0 P_0(\theta_0) L_0(\theta_0, d_0) G}{I_h + I_n + I_o + N} \quad (12.11)$$

where:

- $P_0(\theta_0)$ = home base station (sector 0) overhead ERP including pilot, paging, and sync powers in the direction θ_0 to the probe mobile. Note that, in general, because ERP depends on the antenna pattern (which is a function of direction θ_0), ERP itself is also a function of direction θ_0.

- α_0 = fraction of home base station overhead ERP allocated to pilot power.

- $L_0(\theta_0, d_0)$ = path loss from the home base station in the direction θ_0 to the probe mobile a distance d_0 away.

- G = receive antenna gain of probe mobile.

- I_h = power received at the probe mobile from overhead power emitted by home base station.

- I_n = power received at the probe mobile from other interference of non-CDMA origins. This term is included to accommodate all other possible interference sources that could be jamming the system in the CDMA band.

- I_o = power received at the probe mobile from overhead powers emitted by other base stations.

- N = thermal noise power.

One can define an E_c/I_0 threshold for a particular base station (or a sector of a base station) below which a mobile would no longer consider that sector an active server. Field measurements and simulation have shown that, on average, an E_c/I_0 threshold of -12 dB provides a sufficient balance of forward- and reverse-link and soft hand-off performance when a mobile transitions from one cell to the next. Also note that the traffic channel powers from either the home base station or other base stations are not included in the denominator of (12.11). In reality, because the spread traffic channel signals are transmitted in the same band, the traffic powers contribute to the interference factor in the denominator. The more mobiles there are in the vicinity, the more traffic channel interference is present in the denominator, and the smaller the received E_c/I_0 at a particular probe mobile. Therefore, with a given E_c/I_0 threshold, the effective cell forward service area expands and shrinks depending on traffic loading. This phenomenon is called *cell breathing*. The extent of cell breathing for a particular base station varies throughout the day and depends on the traffic pattern in the area. The unloaded case characterized by (12.11) is used for the purpose of SAB determination; note that this case provides a conservative estimate of forward SAB from the perspective of interference, subject to the constraint of the reverse link. In the unloaded case,

$$I_b = P_0(\theta_0)L_0(\theta_0, d_0)G \qquad (12.12)$$

However, the interference factor I_o also includes terms contributed from other cells or sectors in the vicinity:

$$I_o = G\sum_{k=1}^{K} P_k(\theta_k)L_k(\theta_k, d_k) \qquad (12.13)$$

where $L_k(\theta_k, d_k)$ is the path loss from the kth neighboring cell (or sector) to the probe mobile; the value K is the number of neighboring cells or sectors whose transmitted powers sufficiently contribute to the interference perceived by the probe mobile, and the summation is over this group of neighboring interfering cells or sectors.

Solving (12.11) for $L_0(\theta_0, d_0)$ yields

$$L_0(\theta_0, d_0) = \frac{\dfrac{E_c}{I_0}\left(G\sum_{k=1}^{K} P_k(\theta_k)L_k(\theta_k, d_k) + I_n + N\right)}{\alpha_0 P_0(\theta_0)\left(1 - \dfrac{E_c}{I_0\alpha_0}\right)G} \qquad (12.14)$$

Similarly, the energy per bit per noise power density (E_b/N_0) for the reverse traffic link in the unloaded case is

$$\frac{E_b}{N_0} = \frac{T'L'_0(\theta_0, d_0)G_0(\theta_0)}{I'_n + N}\left(\frac{W}{R}\right) \qquad (12.15)$$

where

- T' = reverse traffic channel ERP of the probe mobile; the transmit pattern is assumed to be omnidirectional.

- $L'_0(\theta_0, d_0)$ = reverse path loss from the probe mobile in the direction θ_0 to the home base station a distance d_0 away.
- $G_0(\theta_0)$ = receive antenna gain of home base station in the direction θ_0 to the probe mobile
- I'_n = power received at the probe mobile from other interference of non-CDMA origins. This term is included to accommodate all other possible interference sources that could be jamming the system in the CDMA band.
- N = thermal noise power.
- (W/R) = processing gain.

As in the case of forward link, one can define an E_b/N_0 threshold for the probe mobile below which it would no longer sustain a satisfactory reverse link. Field measurements and simulation have shown that, on average, an E_b/N_0 threshold of 9 dB is a sufficient target E_b/N_0. Also note that one of the features of a CDMA system is soft handoff, where signals received at two separate base stations (or sectors) from the same mobile are combined; that is,

$$\left(E_b/N_0\right)_{\text{eff}} = f\left[\left(E_b/N_0\right)_0, \left(E_b/N_0\right)_1, \left(E_b/N_0\right)_2\right] \qquad (12.16)$$

where $(E_b/N_0)_k$ = energy per bit to noise power density ratio received at the kth base station (or sector). The 0th base station (or sector) is by definition the best server.

Although the cost of this approach is additional channel element resource at the base stations, the quality of the reverse link is improved through this diversity technique. However, the soft-handoff gain is dependent on the specific network involved and the distribution of cells in the vicinity. In the case of SAB determination, we used the best server's E_b/N_0 as defined in (12.15). Solving (12.15) for $L'_0(\theta_0, d_0)$ yields

$$L'_0\left(\theta_0, d_0\right) = \frac{\dfrac{E_b}{N_0}\left(I'_n + N\right)}{T'G_0(\theta_0)} \frac{1}{\left(W/R\right)} \qquad (12.17)$$

Given a particular cell antenna height h and for a particular radial θ_0, the forward and reverse path losses $L_0(\theta_0, d_0)$ and $L'_0(\theta_0, d_0)$ are functions of distance d_0. There are many propagation models that are used to predict path loss (e.g., Lee, Hata, and others). We invoke the Lee model [5]; for a particular radial d and cell antenna height h, the path loss (in decibels) as a function of distance d is

$$L(d) = -147.7 - 38.4 \log(d) + 20 \log(h)$$
$$\text{and } L(d) = L'(d) \tag{12.18a}$$

where d is in miles and h is in feet. Converting (12.18a) into SI units, we obtain

$$L(d) = -129.45 - 38.4 \log(d) + 20 \log(h) \tag{12.18b}$$

where d is in kilometers and h is in meters. Converting (12.18b) to linear yields

$$L(d) = 1.14 \times 10^{-13} d^{-3.84} h^2 \tag{12.18c}$$

where d is in kilometers and h is in meters. Note that (12.18c) is dimensionless.
Substituting (12.18c) into (12.14), we obtain

$$d_f = 4.26 \times 10^{-4} h^{0.521} \left(\frac{\frac{E_c}{I_0} \left\{ G \sum_{k=1}^{K} P_k(\theta_k) L_k(\theta_k, d_k) + I_n + N \right\}}{\alpha_0 P_0(\theta_0) \left(1 - \frac{E_c}{I_0 \alpha_0} G \right)} \right)^{-0.26}$$

in km
$$\tag{12.19}$$

Substituting (12.18c) into (12.17) yields

$$d_r = 4.26 \times 10^{-4} h^{0.521} \left(\frac{\frac{E_b}{N_0} (I'_n + N)}{T' G_0(\theta_0)} \frac{1}{(W/R)} \right)^{-0.26} \quad \text{in km}$$

$$\tag{12.20}$$

12.2.3 CDMA SAB Determination With Single Sector

Although (12.19) is closed form in nature, it requires the summation over K interfering cells or sectors; this summation is influenced greatly by the physical configurations and parameters of the neighboring cells, and in some extreme cases of cells that are far away. The FCC stated in [6] that,

"We believe that the use of…formula will simplify and remove a measure of uncertainty from the process of calculating and plotting CGSAs…we hope to avoid generally unproductive disagreements between parties and the commission over whether one method of ascertaining the [SAB] is more or less 'accurate' than another for a particular location or terrain."

In developing a standardized method of calculating CDMA SAB, it is desirable to have a standardized formula that is readily applicable and at the same time produces realistic SAB distances. Here we derive the SAB distance based on a single-cell environment.

The ratio of pilot chip energy to total received power spectral density (E_c/I_0) at the probe mobile in a single-cell environment is

$$\frac{E_c}{I_0} = \frac{\alpha_0 P_0(\theta_0) L_0(\theta_0, d_0) G}{I_b + I_n + N} \tag{12.21}$$

Solving (12.21) for $L_0(\theta_0, d_0)$ yields

$$L_0(\theta_0, d_0) = \frac{\dfrac{E_c}{I_0}(I_n + N)}{\alpha_0 P_0(\theta_0)\left(1 - \dfrac{E_c}{I_0 \alpha_0}\right) G} \tag{12.22}$$

Substituting (12.18c) into (12.22) and solving for d, we obtain for the forward link SAB distance as a function of radial θ_0:

$$d_f = 4.26 \times 10^{-4} \, h^{0.521} \left(\frac{\dfrac{E_c}{I_0}(I'_n + N)}{\alpha_0 P_0(\theta_0)\left(1 - \dfrac{E_c}{I_0 \alpha_0}\right) G} \right)^{-0.26} \quad \text{in km} \tag{12.23}$$

Since we are only concerned with the unloaded case, the reverse SAB distance is the same as that of (12.20), also a function of radial θ_0:

$$d_r = 4.26 \times 10^{-4} \, h^{0.521} \left(\frac{\dfrac{E_b}{N_0}(I'_n + N)}{T' G_0(\theta_0)} \frac{1}{(W/R)} \right)^{-0.26} \quad \text{in km}$$

We can further simplify (12.23) using assumptions outlined in Sections 12.2.1 and 12.2.2. For an (E_c/I_0) of -12 dB and a total noise power of -99 dBm, and taking the mobile receive antenna gain to be -2 dB, we obtain

$$d_f = 1.75h^{0.521}\left(P_\alpha - 0.063P_0\right)^{0.26} \tag{12.24}$$

where h is the antenna height in meters, P_α is the base station pilot ERP in watts, and P_0 is the base station total overhead ERP in watts.

We can simplify (12.20) in the same manner. For an (E_b/N_0) of 9 dB and a vocoder rate of 9.6 Kbps, and assuming that the portable eventually transmits at an ERP maximum of 20 dBm to close the reverse link, we obtain

$$d_r = 1.088h^{0.521}G_0^{\,0.26} \tag{12.25}$$

where G_0 is the receive antenna gain of the home base station for the radial in question.

Figure 12.2 depicts the comparison of forward service distances between AMPS characterized by (12.10) and CDMA characterized by (12.24). The

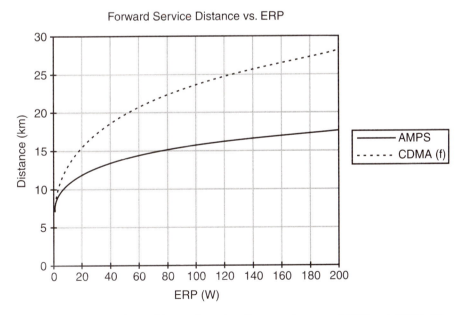

Figure 12.2 A comparison of forward service distances between AMPS and CDMA. The antenna height is 70 ft for both cases.

pilot ERP is assumed to be about 50% of all CDMA overhead power. From the perspective of forward-link coverage, CDMA carries a greater range than AMPS for a given power expenditure. As an example, AMPS needs to transmit 20W ERP in order to attain a forward service distance of 12 km, while for the same amount of overhead power, CDMA can attain a forward service distance of 15.5 km.

Figure 12.3 shows the CDMA reverse service distance as a function of base station receive antenna gain. As expected, service distance is extended as the receive antenna gain increases. However, note that the range of reverse service distance is substantially lower than that of forward service distance, indicating the reverse-link limited nature of CDMA.

12.2.4 CDMA SAB Determination With Power Spectral Density

For any definition of a digital SAB, interference should also be considered since CGSA can be thought of as the area within which the carrier is entitled to protection from interference. The total measured power within a CDMA carrier band cannot be compared directly with the power of an AMPS carrier because for CDMA, the power is spread across the 1.25-MHz band, while the AMPS power is distributed within the 30-kHz band.

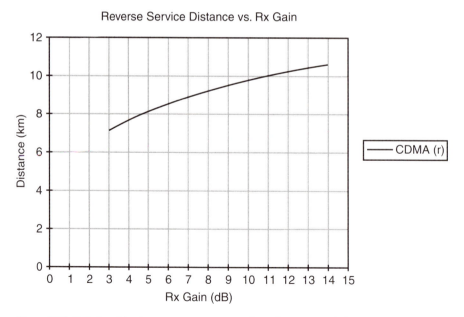

Figure 12.3 Relationship between reverse service distance and base station receive antenna gain. The antenna height is 70 ft.

Let P_c be the deterministic overhead power (ERP) of a CDMA base station. The power is distributed across the entire 1.25-MHz band. Assuming that the power is uniformly spread across the band, the power spectral density can be defined as

$$G_c(f) = \frac{P_c}{1.25 \times 10^6} \text{ W / Hz} \qquad \text{for } f_0 < f < f_1; \ (f_1 - f_0) = 1.25 \text{ MHz}$$

$$0 \qquad \text{everywhere else} \qquad (12.26)$$

Now, let $H_a(f)$ be the transfer function of a perfectly band-limiting unity filter between f_2 and f_3 and $(f_3 - f_2) = 30$ kHz. Then, the interference power radiated to an AMPS mobile in the 30-kHz band from a CDMA base station is

$$P_i = \int_{-\infty}^{+\infty} |H_a(f)|^2 G_c(f) df \qquad (12.27)$$

Given that $H_a(f)$ is unity and perfectly band-limiting, substituting (12.26) into (12.27) yields

$$P_i = \int_{f_2}^{f_3} \frac{P_c}{1.25 \times 10^6} df = P_c \frac{30 \times 10^3}{1.25 \times 10^6} = 0.024 P_c \qquad (12.28)$$

The bandwidth of P_i is 30 kHz. Therefore, under unloaded conditions, the cochannel interference power being radiated to an AMPS mobile in a neighboring cellular system is P_i. Substituting the result of (12.28) into (12.10), we obtain

$$d = 2.531 b^{0.34} (0.024 P_c)^{0.17} = 134 b^{0.34} P_c^{0.17} \qquad (12.29)$$

Figure 12.4 shows the forward service distance calculated using the power spectral density method. As expected, the service distance increases monotonically with increasing overhead ERP. The reverse service distance is also shown for comparison purposes. One curve shows coverage from a coverage perspective (i.e., CDMA (r)), while the other curve shows coverage from an interference perspective (i.e., CDMA (psd)). Note that reverse service distance is independent of base station overhead ERP. Comparing Figure 12.4 with

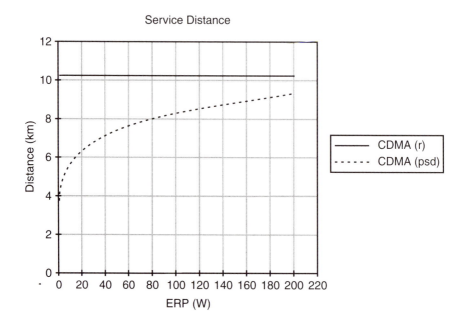

Figure 12.4 Comparison of service distances calculated using power spectral density and reverse link budget. The antenna height is 70 ft for both cases. As an example, the reverse link uses a receive antenna gain of 12 dB, effectively a 13-dB antenna with 1 dB of line loss.

Figure 12.2, we also find that from the point of interference, CDMA has a smaller range than its AMPS counterpart [7].

12.3   RF Exposure Rules

The *National Environmental Policy Act* (NEPA) of 1969 mandates federal government agencies such as the FCC to evaluate the effects of their actions on the quality of the human environment. To comply with NEPA, the FCC rules on human exposure to RF energy emitted by FCC-regulated transmitters and facilities, such as cellular and PCS base stations. The FCC first adopted guidelines in 1985 to evaluate human exposure to RF emissions, and in 1996, the FCC revised and updated these guidelines. These guidelines incorporate limits for MPE in terms of electric and magnetic field strength and power density for transmitters operating at frequencies between 300 kHz and 100 GHz. The FCC's MPE limits are based on recommendations of the *National Council on Radiation Protection and Measurements* (NCRP). These limits are also generally based on guidelines developed by the *Institute of Electrical and Electronics*

Engineers (IEEE) and adopted by the *American National Standards Institute* (ANSI) [8].

12.3.1 Maximum MPE Limits

The FCC issues guidelines and procedures for evaluating the environmental effects of RF emissions. The guidelines generally involve two tiers of exposure limits. The first tier is based on whether exposure occurs in an occupational or "controlled" situation, and the second tier is based on whether exposure could occur to the general population in an "uncontrolled" situation. In general, the exposure limit is five times more stringent for the general public than it is for occupational situations.

Occupational/controlled exposure involves those situations where people are exposed to RF emission as a result of their employment and where those people have been made fully aware of the potential for exposure. This type of exposure also applies to situations where exposure is a consequence of transient passage through a location where exposure levels may be above general population/uncontrolled limits; as such, the exposed person can exercise control over his or her exposure by leaving the area.

General population/uncontrolled exposure involves those situations where the general public may be exposed to RF radiation. This type of exposure also applies to those people who are exposed as a consequence of their employment and who may not have been made aware of the potential for exposure. The general public always falls under this category when exposure is not employment-related. One example could be residents in the vicinity of a cellular or PCS base station.

Table 12.1 shows the limits for maximum permissible exposure in the occupational/controlled situation, and Table 12.2 shows the same limits for the general population/uncontrolled exposure [8]. For both controlled and uncontrolled situations, the MPE limits are specified in units of mW/cm^2 and are different at different frequencies.

As shown in Tables 12.1 and 12.2, the time of exposure is also an important consideration. The averaging time for occupational/controlled exposure is six minutes, and the averaging time for general population/uncontrolled exposure is thirty minutes. For example, a PCS engineer who is fully aware of his or her potential for exposure could be exposed to a power density of 10 mW/cm^2 for three minutes during any six-minute period as long as he or she was exposed at a zero level for the preceding or following three minutes. In general, (12.30) applies:

$$S_e t_e = S_l t_a \tag{12.30}$$

Table 12.1

Limits for Occupational/Controlled Exposure [8]

Frequency Range (MHz)	Power Density (mW/cm^2)	Averaging Time (min)
0.3–3.0	100	6
3.0–30	$(900/f^2)$	6
30–300	1.0	6
300–1,500	$f/300$	6
1,500–100,000	5	6

Table 12.2

Limits for General Population/Uncontrolled Exposure [8]

Frequency Range (MHz)	Power Density (mW/cm^2)	Averaging Time (min)
0.3–1.34	100	30
1.34–30	$(180/f^2)$	30
30–300	0.2	30
300–1,500	$f/1,500$	30
1,500–100,000	1.0	30

where:

- S_e = power density level of exposure in mW/cm^2;
- t_e = allowable time of exposure for S_e;
- S_l = specified power density MPE limit in mW/cm^2;
- t_a = specified MPE averaging time.

12.3.2 Application of MPE Limits

The exposure limits outlined in the previous section are generally applicable to all facilities, operations, and transmitters regulated by the FCC. However, at the time of this writing, the FCC had exempted from routine evaluation cellular towers with transmitting antennas mounted higher than 10m above ground; the FCC has also exempted cellular rooftop facilities with a total power (of all channels) less than 1,000W ERP. For PCS systems, the FCC has exempted from routine evaluation PCS towers with transmitting antennas mounted higher than 10m above ground; the exemption also applies to PCS rooftop facilities with a total power (of all channels) less than 2,000W ERP. Table 12.3 summarizes the MPE evaluation criteria [9].

Given the evaluation criteria outlined in Table 12.3, it is necessary to determine whether or not the total power of all channels from a particular cell site exceeds 1,000W ERP for cellular or 2,000W ERP for PCS. Note that if multiple service providers (e.g., cellular and PCS) collocate at the same site, then the total power of *all* transmitters needs to be considered when performing the calculation. By definition, the total power of all channels is the summation of ERP of all operating transmitters in the facility. For sites using sectorized transmitting antennas, we only need to consider the maximum possible ERP in a single sector (i.e., maximum ERP in the boresight of the sectorized antenna).

For CDMA systems, the calculation of "total power of all channels" is complicated by the fact that the system uses forward power control to maximize coverage and capacity. CDMA's forward transmit power on the traffic channel is continuously changing in response to changing link conditions. The power of any single CDMA traffic channel varies between its maximum and minimum levels. As such, the power of a single channel, as well as the total power of all channels transmitted by a CDMA base station, is statistical rather than deterministic in nature. Furthermore, it is unlikely that all channels operate at their maximum levels at all times. Given that the actual traffic power is statistical in nature, some assumptions could be applied to calculate the total power of all channels. Assuming that the traffic channel power is uniformly distributed between its maximum and minimum levels, the average traffic channel power could be used to calculate the total power of all channels.

Table 12.3
MPE Evaluation Criteria for Cellular and PCS Systems [9]

Service	Facility	Criteria for MPE Evaluation
Cellular	Non-rooftop antennas	Height above ground level to radiation center < 10m and (total power of all channels) > 1,000W ERP
Cellular	Rooftop antennas	(Total power of all channels) > 1,000W ERP
PCS	Non-rooftop antennas	Height above ground level to radiation center < 10m and (total power of all channels) > 2,000W ERP
PCS	Rooftop antennas	(Total power of all channels) > 2,000W ERP

To calculate the average ERP of one traffic channel, $\overline{T}$, we use

$$\overline{T} = \bar{t}lG \qquad (12.31)$$

and

$$\bar{t} = \frac{t_{max} - t_{min}}{2} + t_{min} \qquad (12.32)$$

where

- t_{max} = maximum transmitter output power of one traffic channel;
- t_{min} = minimum transmitter output power of one traffic channel;
- $\bar{t}$ = average transmitter output power of one traffic channel;
- l = transmission line loss (a number between 0 and 1) between transmitter output and antenna;
- G = transmit antenna gain at boresight, or maximum antenna gain.

To calculate the total power (i.e., ERP) of all channels, P_{TOTAL}, of a specific base station, we use the average traffic channel power along with deterministic overhead powers (i.e., pilot, paging, and sync) to calculate P_{TOTAL}:

$$P_{TOTAL} = P_\alpha + P_\beta + P_\gamma + N\overline{T} \qquad (12.33)$$

where

- P_α = pilot ERP;
- P_β = paging ERP;
- P_γ = sync ERP;
- N = number of traffic channels equipped in the base station.

Another (conservative) way of calculating P_{TOTAL} is to use the maximum power of the power amplifier as the transmitter output power. In other words, P_{TOTAL} is

$$P_{TOTAL} = plG$$

where p is the transmitter output power (i.e., maximum power of the power amplifier) of the base station.

The value P_{TOTAL} should be calculated for every base station in the system to determine whether or not they meet the MPE evaluation criteria summarized in Table 12.3. If P_{TOTAL} for a cellular rooftop facility or a cellular tower facility less than 10m in height is greater than 1,000W, or if P_{TOTAL} for a PCS rooftop facility or a PCS tower facility less than 10m in height is greater than 2,000W, then a determination of compliance with the exposure limits will be necessary. In the following section, we outline some commonly acceptable methods of calculating power densities. The calculated power densities then can be compared with the specified MPE limits summarized in Tables 12.1 and 12.2 to determine compliance.

Note that the two calculation methods just presented are only proposals and have not been approved by the FCC. It is the responsibility of each service provider to accurately determine the realistic total power and ERP. That determination may involve more elaborate calculation methods or actual measurements.

12.3.3 Evaluation of MPE Power Densities

Many methods exist to model RF field strength and power density levels around radiating antennas. The simplest method of predicting power density levels in the far field of an antenna is summarized by (12.34):

$$S(\theta, \phi) = \frac{1.64 P(\theta, \phi)}{4 \pi r^2} \qquad (12.34)$$

where

- S = power density;
- P = antenna ERP as a function of horizontal direction θ and vertical direction ϕ from the radiating antenna;
- r = distance to the radiation center of antenna.

As before, antenna ERP P is calculated using the following equation:

$$P(\theta, \phi) = p l G(\theta, \phi) \qquad (12.35)$$

where

- P = transmitter output power;

- $G(\theta,\phi)$ = antenna gain as a function of horizontal direction θ and vertical direction ϕ from the radiating antenna.

Note that since antenna gain is a function of θ and ϕ and ERP is a function of antenna gain, power density is a function of θ and ϕ. The factor of 1.64 is the gain of a half-wave dipole relative to an isotropic radiator; since ERP is calculated using antenna gain referenced to a half-wave dipole, the factor of 1.64 is multiplied by P to produce, in the numerator, a radiated power quantity that is relative to isotropic.

For a worst-case prediction of power density at or near a surface, such as at the ground level or on a rooftop, we could assume a 100% reflection of incoming radiation, and this assumption results in a doubling of predicted field strength, or a fourfold increase in far-field power density. For this worst-case scenario, (12.34) can be modified to

$$S(\theta,\phi) = \frac{(2)^2 1.64 P(\theta,\phi)}{4\pi r^2} = \frac{1.64 P(\theta,\phi)}{\pi r^2} \qquad (12.36)$$

In situations where the radiating antenna is located on the rooftop and there is a possibility that people may have access to areas within several wavelengths, or the near field, of the antenna, then (12.34) and (12.36) may no longer hold true. In the case of predicting power density at distances close in to the antenna, a model based on inverse distance is more accurate than predictions based on far-field equations such as (12.34) and (12.36). In the near field, the power density decays as a function of r rather than r^2. A simple and accurate model to use is the cylindrical model where the radiated power is distributed over the surface area of an imaginary cylinder [10]. For an omnidirectional antenna,

$$S(\theta,\phi) = \frac{1.64 P(\theta,\phi)}{2\pi r h} \qquad (12.37)$$

where h is the dimension of the antenna. For a sectorized antenna,

$$S(\theta,\phi) = \frac{1.64(360)P(\theta,\phi)}{2\pi r h \, \theta_3} \qquad (12.38)$$

where θ_3 is the 3-dB beam width of the antenna in degrees.

12.3.4 RF Mitigation Measures

If it is determined, either through the models described in the previous section or through actual measurements, that a facility is not compliant with the FCC MPE guidelines, then implementation of mitigation measures is required for that particular facility. There are several mitigation measures in common use.

The first mitigation measure is access restriction. Restricting access is the simplest means of controlling exposure to areas where high RF power density levels may be present. This method includes fencing and locking out unauthorized persons in appropriate areas. When access to the appropriate area is restricted, the facility or operation can certify that it complies with the FCC requirements [8].

The second mitigation measure is duration of access. After determining the power density level of exposure in a particular area, (12.30) can be used to determine the allowable time of exposure, This way, time-averaging aspects of the exposure limits can be used to limit the duration of access to appropriate areas. Understandably, this type of restriction is more practical in occupational and controlled situations.

The third mitigation measure is antenna placement. Engineers should consider MPE guidelines when designing cell sites. In a rooftop situation, transmitting antennas should be elevated to the extent possible to increase r and thus to lower the power density in appropriate areas. One should certainly keep transmitting antennas away from publicly accessible areas. Directional antennas should be located near the edge of the roof and pointed away from the building.

The fourth mitigation measure is shielding and power reduction. Although reducing transmit power can lead to performance impacts in some cases, it should be emphasized that complying with FCC rules carries the highest priority. The transmit power can be reduced; shielding is also an alternative, and both RF shielding and power reduction can be implemented simultaneously to bring a facility into compliance with MPE limits.

12.4 Remarks

In this chapter, we have described three alternative methods of calculating SAB for a CDMA cellular system. The result obtained in a multiple-cell scenario takes into account interference from nearby cells. The result is analytically correct but difficult to calculate, since the equations are functions of not just the periphery cell but also interfering cells. The results obtained in the single-cell scenario are closed form in nature and simplify the final equations to only functions of parameters of the cell site in question. Two functions are derived—forward and reverse—and SAB is arrived at based on both forward-link pilot channel and reverse-link traffic channel coverage boundaries. Commercially deployed CDMA systems demonstrated that the proposed methods provide more accurate estimation of coverage boundary than the existing AMPS-based method. Finally, considering interference protection and maintaining continuity from the current AMPS SAB formula, we derived a single equation using the power spectral density of both AMPS and CDMA carriers.

In terms of the FCC's RF exposure rules, we have described the FCC guidelines and methods of determining whether or not a facility is compliant with relevant FCC rules. It is important to note, however, that the FCC rules are subject to change, and it is important to be compliant with the current rules that are in force.

References

[1] *Code of Federal Regulations, Telecommunications.* Part 22 Public Mobile Service, ss 22.911, 1995.

[2] Carey, R. B., "Technical Factors Affecting the Assignment of Facilities in the Domestic Public Land Mobile Radio Service," *FCC Report No. R-6406,* June 1964.

[3] Lee, W. C. Y., "Comments of PacTel Cellular in the Matter of Amendment of Part 22," *FCC Docket No. 90-6,* Dec. 1991.

[4] Lee, W. C. Y., *Mobile Cellular Telecommunications: Analog and Digital Systems,* New York, NY: McGraw Hill, 1995.

[5] Propagation Ad Hoc Committee, "Lee's Model," *IEEE Trans. on Vehicular Technology,* Feb. 1988, Special Issue.

[6] "FCC Second Report and Order, In the Matter of Amendment of Part 22 of the Commission's Rules to Provide for Filing and Processing of Applications for Unserved Areas in the Cellular Service and to Modify other Cellular Rules," *FCC 92-94:38357,* April 1992.

[7] Yang, S. C., A. Watson, and J. Yang, "CGSA Determination in a CDMA Cellular System," *Proc. 9th Annual International Conf. on Wireless Communications,* Calgary, Alberta, Canada, July 9-11, 1997, pp. 452–464.

[8] Cleveland, R. F., D. M. Sylvar, and J. L. Ulcek, "Evaluating Compliance with FCC-Specified Guidelines for Human Exposure to Radio Frequency Radiation," *FCC Office of Engineering and Technology Bulletin No. 65,* 1996.

[9] "FCC Report and Order on Guidelines for Evaluating the Environmental Effects of Radio Frequency Radiation," *FCC ET Docket No. 93-62,* Aug. 1996.

[10] Gailey, P. C., and R. A. Tell, "An Engineering Assessment of the Potential Impact of Federal Radiation Protection Guidance on the AM, FM, and TV Broadcast Services," *U.S. Environmental Protection Agency Report No. EPA 520/6-85-011,* April 1985.

About the Author

Samuel C. Yang holds two graduate degrees from Stanford University and an undergraduate degree from Cornell University, all in electrical engineering. He is currently a manager of the system engineering group at AirTouch Cellular - Southern California, where he played a key role in the design and commercialization of the first large-scale CDMA system in the United States. Samuel Yang is a registered professional engineer in the state of California.

Prior to coming to AirTouch in 1995, Samuel Yang worked at Hughes Space and Communications Company, where he served as a technical lead on several international satellite projects for China, Japan, and Thailand. While at Hughes, he also conducted research in advanced multiple-access techniques and channel simulation for mobile satellite communications, as well as served as a system engineer on NASA's Magellan radar-mapping mission to Venus. His current interests are system planning, design, and optimization of mobile wireless networks.

Index

Access channel, 119–22
Access channel frame, 121
Access channel message, 121–22
Access channel slot, 121
Access parameters message, 88, 112–13,
    141, 137, 139
Access probe, 85–88
Access procedures, 141–45
Access state, 134, 139–45
Active set, 99, 100–2, 137, 184
Adaptive pulse code modulation, 32
A/D converter. *See* Digital-to-analog
    converter
Additive white Gaussian noise, 62, 72
ADPCM. *See* Adaptive pulse code
    modulation
Advanced mobile phone
    system, 10, 31, 211, 213, 215,
    217–18, 249–53, 259–61
Alert with information message, 146
Algorithmic decision-making, 238–39
All-pole filter, 34–36
All-zero filter, 35
AM. *See* Amplitude modulation
American National Standards
    Institute, 263
Amplifier, linear and nonlinear, 210
Amplitude, signal pulse, 29–30
Amplitude modulation, 58

AMPS. *See* Advanced mobile phone
    system
Analog modulation, 58
Analysis filter, 35
Analysis function, 246
Analytical methods, 245
ANSI. *See* American National Standards
    Institute
Antenna aperture, 251–52
Antenna configuration, 58, 192–93,
    216, 269
Antenna temperature, 201
Application-specific integrated circuit, 31
ASIC. *See* Application-specific integrated
    circuit
Asymmetric link, 46, 105
Authentication challenge message, 141
Authentication challenge response
    message, 140–41
Autocorrelation, 56–57
AWGN. *See* Additive white Gaussian
    noise

Background noise, 94
Bandpass filter, 211, 213
Bandwidth expansion factor, 5
Baseband filter, 106
Base station, 85–87, 89–90, 94, 99,
    120–21, 124, 192, 217–18, 253

B-CDMA. *See* Broadband code division multiple access
BER. *See* Bit error rate
Billing function, 246
Binary phase-shift keying, 58–65
Bit error rate, 65, 75, 156
Bit interval, 5
Blank and burst technique, 127
Block code. *See* Linear block code
Blocking, call
    hard, 225, 231–33
    soft, 225–31
Blocking probability, 217, 220
Blocking rate, 217, 220, 238
Boltzmann's constant, 14, 198, 250
BPSK. *See* Binary phase-shift keying
Broadband code division multiple access, 25
Burst error, 43

Call accounting, 245–46
Call detail record, 245
Call processing, 133–35
    access state, 139–45
    idle state, 136–39
    initialization state, 135–36
    traffic channel state, 145–47
Candidate set, 99, 100, 102, 184
Capacity, system, 30–31, 75–78, 179
    loading effects, 78–79
    low-noise amplifier, 208
    sectorization, 79–82
    voice activity, 82–83
See also Power control
Carried load, 220
Carrier-to-interference ratio, 14
Carrier-to-noise ratio, 13–14
CDMA. *See* Code division multiple access
Cell breathing, 255
Cellular geographic service area, 249, 260
CELP. *See* Code-excited linear prediction
CGSA. *See* Cellular geographic service area
Channel assignment message, 140
Channel coding, 36–37
    convolutional, 41–42, 89, 114, 117, 120, 125, 177
    interleaving, 43

linear block, 37–41
Channel decode function, 32
Channel encode function, 31
Channelization
    pseudo noise code, 53–55
    Walsh code, 48–50
Channel list message, 137
Channel supervision, 181–82
Charging function, 245
Chip interval, 5, 49, 121, 165–66, 168
C/I. *See* Carrier-to-interference ratio
Closed-loop power control, 88–94
C/N. *See* Carrier-to-noise ratio
Code division multiple access, 1–2
    See also IS-95 CDMA system
Code-excited linear prediction, 35–36
Coherent demodulation, 69
Coherent link, 118
Collision, access channel, 141
Complementary error function, 64
Composite noise temperature, 200–1
Configuration management, 243–44
Configuration messsage, 138
Congestion, mobile communication, 141
Conversation substate, 147
Convolutional code, 37, 41–42, 89, 114, 117, 120, 125, 177
Correlation crosstalk, 9
Correlator demodulator, 69
Coverage, mobile area, 176–79, 189, 192
CRC. *See* Cyclic redundancy check
Cross-correlation, 2–3, 47
Cross-product, 2
Cyclic redundancy check, 39–41, 125, 181

Database, 237
Data burst message, 141
Data burst randomizer, 123–25
Decision support system, 239
Decision threshold, 50, 55, 69
Default constant, 87
Deinterleaver, 43
Delay spread, 23–26
Delta modulation, 32
Demodulator/demodulation, 32
    binary phase-shift keying, 59–61
    quadrature phase-shift keying, 67–69

Demultiplexer, 67
Digital communication
    advantages, 29–31
    system components, 31–32
Digital signal processing, 31
Digital-to-analog converter, 5
Dim and burst technique, 128
Direct-sequence spread spectrum, 2, 75
    applications, 9–11
    multiple access, 2–9
Walsh code, 47–50
Discrete-time autocorrelation, 56
Diversity technique, 256
DM. *See* Delta modulation
Doppler effect, 20–21
Dot product, 3
DSP. *See* Digital signal processing
DSS. *See* Decision support system
DS-SS. *See* Direct-sequence spread
        spectrum
Dual mode, 135

Effective noise power, 14
Effective radiated power, 13, 87, 154,
        156, 163, 190–91, 194, 216,
        261, 264–68
Electromagnetic energy, 250
Electromagnetic field, 58
Electronic serial number, 117
EME. *See* Electromagnetic energy
EM field. *See* Electromagnetic field
Encryption, 31
Energy per bit per noise power
        density, 76, 156–63, 255
Energy per chip per interference
        density, 149–56
Equivalent circuit, 199
Erasure indicator bit, 176
Erlang, 218–19
Erlang-B model, 221–23
Erlang-C model, 223–24
ERP. *See* Effective radiated power
Error, voice coding, 35
Error-correcting code, 36–43, 72–73
Error protection, 90–91, 114, 120
ESN. *See* Electronic serial number
Extended system parameters message, 137

Fade timer, 182
Fading, 43, 73
    fast, 22, 88–89, 91
    multipath, 19–22
    slow, 19
Far-field power density, 268
Fault management, 240–41
FCC. *See* Federal Communications
        Commission
FDMA. *See* Frequency division multiple
        access
Federal Communications
        Commission, 249–50, 257–58,
        262–64, 267, 269
Feedback control system, 238
Feedback loop, 35
FER. *See* Frame error rate
FH-SS. *See* Frequency-hopping spread
        spectrum
Field optimization, 189–93
Field strength, service boundary, 249–62
Filtering, 106, 211, 213
Filter modeling, voice coding, 34–36
FM. *See* Frequency modulation
Footprint, coverage, 193, 216
Forward link, 41, 46, 51, 57–58, 72, 88, 90,
        94, 96–98, 102, 105, 118, 120
    channel format, 128–29
    channel supervision, 181–82
    coverage/capcity, 176–79, 190–91
    interference, 191–92, 196–97, 213
    paging channel, 109–13
    pilot channel, 106, 149–56
    power control, 175–76
    service area boundary, 250,
        253–54, 258–62
    sync channel, 106–9
    traffic channel, 114–18, 156–58
    voice activity, 123–24
Fourier transform, 25
Frame, paging channel, 110–11
Frame, traffic channel, 125, 128, 181–82
Frame error rate, 83, 89, 91, 93–94,
        124, 174–75, 177–78,
        190–91, 193, 225
Free-space propagation loss, 15, 18

Frequency division multiple
          access, 1–2, 10, 43, 195
Frequency-hopping spread spectrum, 2
Frequency modulation, 58
Frequency planning, 10
Frequency reuse, 10, 78, 164
Frequency-selective fading, 25–26
Frequency shift, 20–21

Gain, 80, 200–1, 206, 216
Gain control function, 118
Gaussian distribution, 19–20, 63–64
Global service redirection message, 137
Grade of service, 217, 220–24

HAAT. See Height above average terrain
Hadamard matrix, 46, 47, 105
Hamming code/distance, 38–39
Handoff, 94–102, 128
    hard, 98–99
    idle, 137
    pilot search, 102
    process, 100–2
    set maintenance, 99–100
    soft/softer, 93, 96–97, 185–86
Handoff completion message, 101, 130
Handoff direction
          message, 99–102, 129–30
Handoff drop timer expiration value, 99
Hata propagation loss model, 16–18
Height above average terrain , 253
Heuristic decision-making, 238

Idle handoff, 137
Idle state, 134, 136–39
IEEE. See Institute of Electrical and
          Electronics Engineers
IM. See Intermodulation
IMSI. See International mobile station
          identity
In-band interference, 196–7
Independent/identically-distributed
          random variable, 20
Indirect signal path, 15
Initialization state, 134–36
Inner loop, 91–92
In-phase signal, 20, 62, 66–67

Institute of Electrical and Electronics
          Engineers, 262–63
Integrator, symbol, 124
Interference, 9, 21, 75, 77, 79–80, 94,
          115, 160–61, 163, 191–93,
          197, 225, 254, 260–61
    See also Noise
Interleaving, 43, 117, 120
Intermodulation, 208–15
International mobile station identity, 139
Intersymbol interference, 23–24
Inventory function, 243
IS-95 CDMA system, 1–2, 24, 29, 31, 35,
          40–47, 51, 57, 72–73, 75, 94,
          105, 127, 133, 253
ISI. See Intersymbol interference

Jammer, 192–93, 197, 241

Lee propagation loss
          model, 15–16, 18, 256
Linear amplifier, 210
Linear block code, 37–41
Linear dynamic programming, 42
Linear feedback shift, 51–52
Linear filter, 34
Linear-predictive coding, 35
Line-of-sight, 15
Link analysis, 13–14
LNA. See Low-noise amplifier
Load impedance, 198
Load, traffic, 220, 232–33
Loading, 78–80, 164, 189, 208
Logical channel, 118, 165
Log-normal distributed fading, 19
Long pseudo noise code, 57
LOS. See Line-of-sight
Low-noise amplifier, 199
    capacity improvement, 208
    signal-to-noise ratio, 205–7
    system with, 202–5
    system without, 200–2
LPC. See Linear-predictive coding

MAHO. See Mobile-assisted handoff
Majority decoding, 37
Management information systems
    call accounting, 245–46

configuration management, 243–44
decision classes, 238–40
fault management, 240–41
performance management, 242–43
planning, 244–45
roles, 236–38
Masking pattern, 125
Matched-filter method, 59, 67
Maximal-length shift register code, 52
Maximum likelihood detector, 61, 67
Maximum permissible
    exposure, 250, 262–69
Maximum power transfer, 198
Maximum slot cycle, 110
Mean opinion score, 174
Memoryless code, 41
Message response substate, 140
Message sequence number, 138
Metering function, 245
Microcell region, 18–19
Minimum distance, 38–39
MIS. *See* Management information
    systems
Mobile-assisted handoff, 99
Mobile call. *See* Call processing
Mobile communication, 9–11
Mobile station message transmission
    substate, 141
Mobile station order, 140
Mobile station origination attempt
    substate, 140
Mobile switching center, 96
Mobile-to-land call, 133
Mobile-to-mobile call, 133
Mobile transmit power, 208–9
Model base, 237
Modulate function, 31
Modulator/modulation, 58, 118, 120
    binary phase-shift keying, 58–65
    code division multiple access, 72–73
    quadrature phase-shift keying, 66–72
MOS. *See* Mean opinion score
MPE. *See* Maximum permissible exposure
MSC. *See* Mobile switching center
Multipath delay spread, 23–26
Multipath distortion, 10
Multipath Rayleigh fading, 19–22

Multiple access, 31–32, 43–46
    pseudo noise code, 51–58
Walsh code, 46–51
Multiplexer/multiplexing, 69, 127

N=7 frequency reuse, 10
National Council on Radiation Protection
    and Measurements, 262
National Environmental Policy Act, 262
NCRP. *See* National Council on Radiation
    Protection and Measurements
Near-far problem, 7–8, 84
Near-field power density, 268
Neighbor list message, 100,
    112–13, 129, 137
Neighbor set, 99, 137, 187
NEPA. *See* National Environmental
    Policy Act
No dominant server, 193
Noise, 196
    binary phase-shift keying, 61–65
    forward link, 196–97
    intermodulation, 208–15
    mobile number, 215–16
    reverse link, 197–98
    thermal, 198–99
    See also Interference; Low-noise
        amplifier
Noise enhancement, 202, 204–5
Noise variance, 70
Nonlinear amplifier, 208, 210
Nonslotted mode, 136
Null, 10

Offered load, 220
Offset quadrature phase-shift
    keying, 72, 121
Omnidirectional antenna, 268
Open-loop power control, 88, 93–94
OQPSK. *See* Offset quadrature phase-shift
    keying
Origination message, 140–41
Orthogonal coding, 2–9, 45–46, 75
Orthogonal modulator, 120–21
Outer loop, 91–92
Out-of-band noise, 196, 211, 213
Overhead channel, 196

Overhead message, 112

Padding bit, 107, 111, 121–22
Page response message, 140
Page response substate, 140
Paging channel, 105, 109–13, 136, 253
Paging channel message, 112–13, 137–39
Parity bit, 38
Partial correlation, 7–9
Path loss. *See* Propagation loss
PCB. *See* Power-control bit
PCG. *See* Power-control group
PCM. *See* Pulse code modulation
PCS. *See* Personal communication system
Performance management, 242–43
Periodic reporting, 182
Persistence delay, 143
Persistence test, 143
Personal communication system, 1,
        235–37, 239, 249, 264–65, 267
    See also Management information
        systems
Phase-shift keying, 30
Pilot channel, 98–99, 105–6,
        149–56, 253–54
Pilot channel acquisition substate, 135
Pilot detection threshold, 99
Pilot drop threshold, 99
Pilot offset, 168
Pilot pollution, 191
Pilot search, 102
Pilot strength, 100–1, 130, 190
Pitch, 34
PMRM. *See* Power measurement report
        message
PN code. *See* Pseudo noise code
Polynomial representation, 39–40
Power control, 83–85, 118
    closed-loop, 88–93
    fast, 176–77
    forward link, 94, 175–76
    open-loop, 88, 93–94
    parameters, 182–83
    reverse link, 85–94, 124, 192, 206, 208
Power-control bit, 89, 93–94, 114, 117, 124
Power-control group, 90, 117, 125
Power-control parameters message, 129

Power density level, 267–68
Power measurement report
        message, 94, 130, 175–76, 191
Power spectral density, 260–62
Power, total, 265–66
Poynting's theorem, 251
Primary data, 127
Privacy, communication system, 31
Probability density function, 21, 63–64
Processing gain, 5, 77, 82, 176, 178
Procurement function, 243
Propagation loss, 14–19, 190,
        201, 216, 256
Pseudo noise code, 46, 51–58, 83, 98,
        102, 105–10, 117
    adjacent PN offset, 171–74
    aliasing, 169–72
    co-PN offset, 168–71
    long, 109–10, 117, 119, 121
    offset planning, 165
    short, 121, 165–68
Pseudorandom masking, 125
PSK. *See* Phase-shift keying
PSTN. *See* Public switched telephone
        network
Public switched telephone network, 133
Pulse code modulation, 32, 33

Quadrature phase-shift keying, 66–72, 121
Quadrature signal representation, 20, 62
Quality of service, 29–30, 75

Radio frequency channel, 10
Radio frequency exposure, 262–69
Radio frequency mitigation, 269
Rake receiver, 10, 24, 96
Random transmission time, 141, 143
Rayleigh fading, 19–22, 88–89, 91
Received power, 191, 193, 252
Receiving equipment, 200
Recursion, 46
Redundancy bit, 36–38, 40–41
Reflection, signal, 15, 19
Refraction, signal, 15
Registration access substate, 140
Registration message, 140
Release substate, 147

Remaining set, 99, 137
Request message, 122
Response message, 122, 141
Reverse link, 10, 42, 46, 51, 57, 72, 77,
        83, 95–97, 118–19, 158–59
    access channel, 85–88, 119–22
    channel format, 130
    channel supervision, 182
    closed-loop, 88–94
    coverage/capacity, 178–79, 192
    frequency reuse, 164
    interference, 192–93, 197–98, 215
    open-loop, 88, 93–94
    power control, 206, 208
    quality, 89, 93
    rise, 163–64, 201, 206–7
    service area boundary, 250,
        253–57, 260–62
    traffic channel, 122–25, 159–63
RF channel. *See* Radio frequency channel
Rise, reverse-link, 163–64, 201, 206–7
Robbing, bit, 89

SAB. *See* Service area boundary
Satellite communications, 14–15
SCI. *See* Synchronized capsule indicator
Search-window size, 184–89
Secondary data, 127–28
Sectorization, 79–83
Sectorized antenna, 268
Semistructured decision, 238
Service area boundary, 249
AMPS calculation, 250–53
CDMA multiple sector, 253–57
CDMA power spectral density, 260–62
CDMA single sector, 257–60
Set maintenance, 99–100
Shadowing, 19, 88
Shape, signal pulse, 29–30
Signaling data. *See* Secondary data
Signal-space representation, 62, 68
Signal-to-noise ratio, 76–77, 84, 91,
        99, 199–208
Signal transmission rate, 75
Simulation assessment, 245
Slope factor, 18
Slotted mode, 136

SNR. *See* Signal-to-noise ratio
SOM bit. *See* Start-of-message bit
Source coding, 31–36
Source decode function, 32
Source encode function, 31
Speech coding, 33–36
Spread spectrum, 1–9
Start-of-message bit, 108
Step size, 86
Strength measurement message, 101
Structured decision, 238
Symbol detection, 124
Symbol repetition, 114–16, 120
Sync channel, 105–9, 253
Sync channel acquisition
        substate, 135–36
Sync channel frame body, 108
Sync channel message, 106–9, 136
Sync channel message capsule, 107
Sync channel superframe, 107
Synchronization, 51
Synchronized capsule indicator, 110–11
Synthesis filter, 35
System determination substate, 135
System noise temperature, 201
System parameters
        message, 102, 112, 128, 137

TDMA. *See* Time division multiple access
Terrestrial environment path loss, 15, 18
Thermal noise, 14, 94, 197–99
Threshold detector, 30
Threshold reporting, 182
Time division multiple
        access, 2, 10, 29, 43, 195
Timing change substate, 136
Tracking function, 243–44
Traffic channel, 105, 114–18, 265–66
    formats, 125–30
    forward link, 128–29, 156–58,
        191–92, 253–54
    interference, 196–97
    reverse link, 122–25, 130, 159–63, 193,
        215–16, 253
Traffic channel initialization
        substate, 145–46
Traffic channel state, 134, 145–47

Traffic demand, 217
Traffic engineering, 217
Traffic intensity, 218–19
Transfer function, 210
Tree search algorithm, 42
Trend analysis, 242–43

Unstructured decision, 238
Unsynchronized paging channel message
        capsule, 111
Unvoiced sound, 33
Up-conversion, 118
Update overhead information
        substate, 139
Urban environment, 16–17

Variable-rate coding, 31, 82

Voice activity, 10, 82–83, 123
Voice coder (vocoder), 31, 33–36,
        82, 114–15, 117, 123–25, 174–79
Voice quality, 174–75
Voiced sound, 33

Waiting for mobile station answer
        substate, 146
Waiting for order substate, 146
Walsh code, 46–51, 57, 102, 105–7, 110,
        117–18, 120–21, 165
Waveform coding, 32
W/R. *See* Processing gain

Zero crossing, 72

The Artech House Mobile Communications Series

John Walker, Series Editor

Advanced Technology for Road Transport: IVHS and ATT, Ian Catling, editor

An Introduction to GSM, Siegmund M. Redl, Matthias K. Weber, Malcolm W. Oliphant

CDMA for Wireless Personal Communications, Ramjee Prasad

CDMA RF System Engineering, Samuel C. Yang

Cellular Communications: Worldwide Market Development, Garry A. Garrard

Cellular Digital Packet Data, Muthuthamby Sreetharan, Rajiv Kumar

Cellular Mobile Systems Engineering, Saleh Faruque

Cellular Radio: Analog and Digital Systems, Asha Mehrotra

Cellular Radio: Performance Engineering, Asha Mehrotra

Cellular Radio Systems, D. M. Balston, R. C. V. Macario, editors

Digital Beamforming in Wireless Communications, John Litva, Titus Kwok-Yeung Lo

GSM System Engineering, Asha Mehrotra

Handbook of Land-Mobile Radio System Coverage, Garry C. Hess

Introduction to Wireless Local Loop, William Webb

Introduction to Radio Propagation for Fixed and Mobile Communications, John Doble

Land-Mobile Radio System Engineering, Garry C. Hess

Low Earth Orbital Satellites for Personal Communication Networks, Abbas Jamalipour

Mobile Antenna Systems Handbook, K. Fujimoto, J. R. James

Mobile Communications in the U.S. and Europe: Regulation, Technology, and Markets, Michael Paetsch

Mobile Data Communications Systems, Peter Wong, David Britland

Mobile Information Systems, John Walker, editor

Personal Communications Networks, Alan David Hadden

RF and Microwave Circuit Design for Wireless Communications, Lawrence E. Larson, editor

Smart Highways, Smart Cars, Richard Whelan

Spread Spectrum CDMA Systems for Wireless Communications, Savo G. Glisic, Branka Vucetic

Transport in Europe, Christian Gerondeau

Understanding Cellular Radio, William Webb

Understanding GPS: Principles and Applications, Elliott D. Kaplan, editor

Vehicle Location and Navigation Systems, Yilin Zhao

Wireless Communications for Intelligent Transportation Systems, Scott D. Elliott, Daniel J. Dailey

Wireless Communications in Developing Countries: Cellular and Satellite Systems, Rachael E. Schwartz

Wireless Data Networking, Nathan J. Muller

Wireless: The Revolution in Personal Telecommunications, Ira Brodsky

For further information on these and other Artech House titles, including previously considered out-of-print books now available through our In-Print-Forever™ (IPF™) program, contact:

Artech House
685 Canton Street
Norwood, MA 02062
781-769-9750
Fax: 781-769-6334
Telex: 951-659
e-mail: artech@artech-house.com

Artech House
Portland House, Stag Place
London SW1E 5XA England
+44 (0) 171-973-8077
Fax: +44 (0) 171-630-0166
Telex: 951-659
e-mail: artech-uk@artech.house.com

Find us on the World Wide Web at: www.artech-house.com